HISTORICAL RECORDS

OF THE

XIII MADRAS INFANTRY

HISTORICAL RECORDS
OF THE
XIII MADRAS INFANTRY

COMPILED BY
LIEUTENANT R. P. JACKSON

The Naval & Military Press Ltd

Published by
The Naval & Military Press Ltd
Unit 10, Ridgewood Industrial Park,
Uckfield, East Sussex,
TN22 5QE England
Tel: +44 (0) 1825 749494
Fax: +44 (0) 1825 765701
www.naval-military-press.com

DEDICATED

TO

LIEUTENANT-COLONEL HERBERT LAWSON

COMMANDANT 13TH MADRAS INFANTRY

WHOSE KINDLY INTEREST IN EVERYTHING

WHICH RELATES TO THE REGIMENT UNDER

HIS COMMAND IS SO MUCH APPRECIATED

BY ALL RANKS

“ ’Twere a concealment,
Worse than a theft, no less than a traducement,
To hide your doings.”

SHAKESPEARE.

PREFACE

THESE Records were originally commenced by the Commandant, Lieutenant-Colonel H. Lawson, who, finding that he was unable to spare the time necessary for the deep research into official and other publications their preparation entailed, asked me to take the work over from him. The fact of having to consult and study so many works of reference to make the History complete has rendered my task a pleasant and interesting one, though the labour of compiling the supplementary parts has been far from light, as the manuscript record maintained in the regiment was only apparently written up in 1845, and necessarily was most incomplete, and the statistics have been now tabulated for the first time. It is hoped that at some future date the historical portion will be translated into the vernaculars and read in the regimental school, and thus strengthen the *esprit de corps*, which is so valuable in a regiment. Should these Records fulfil this purpose, and prove of some small interest to those who have served, are serving, or will hereafter serve in the regiment, I shall consider my time and labour have not been thrown away.

Much of the information here reproduced is necessarily of a general nature, and applies equally to all Madras Infantry Regiments, but the history of any particular one would be incomplete without it.

R. P. JACKSON, *Lieutenant,*
Adjutant 13*th* *M.I.*

54 Parliament Street,
London, 21*st* *October* 1898.

CONTENTS

PART		PAGE
I.	Regimental History	15
II.	Establishment and Recruiting	161
III.	Dress, Clothing, and Necessaries	197
IV.	Devices, Badges, and Marks of Distinction	231
V.	Colours	237
VI.	(1) Arms, Ammunition, and Accoutrements	247
	(2) Table of Results of Annual Courses, etc.	259
	(3) Shooting Prizes	261
VII.	(1) List of Inspecting Officers	265
	(2) Succession of British Regimental Staff Officers	267
	(3) List of British Regimental Officers	272
	(4) List of Officers Killed and Wounded	289
	(5) List of Officers and Others who have Distinguished Themselves	290
	(6) List of Native Officers	293
	(7) List of Recipients of the Order of British India	296
	(8) List of Recipients of Good Conduct Medals	297
	(9) Return of Efficiency of Signallers	299
	(10) Return of Certificates of Education	300
	(11) Medical Statistics	301
VIII.	Stations of the Regiment	303
IX.	Miscellaneous—Pay, Pension, etc.	309

LIST OF PLATES

PLATE

I. NATIVE OFFICERS, NON-COMMISSIONED OFFICERS, AND SEPOYS OF MADRAS ARTILLERY AND INFANTRY, 1791 (*see page* 201).

II. SUBADAR, GRENADIER, AND RECRUIT OF MADRAS INFANTRY, 1799 (*see page* 201).

III. BRITISH OFFICER, 1776 (*see page* 200).
SEPOY, 1806 (*see page* 204).

IV. BRITISH OFFICER, 1816 (*see page* 207).

V. SEPOY, 1806 (*see page* 204).
SEPOY, 1824 (*see page* 209).
BANDSMAN, 1840 (*see page* 210).
SUBADAR, 1850 (*see page* 210).

VI. SEPOY (Drill Order), 1860 (*see page* 213).
SEPOY (Review Order), 1860 (*see page* 213).
SEPOY (Drill Order), 1861 (*see page* 213).
RECRUIT BOY, 1866 (*see page* 213).

VII. SEPOY (Drill Order), 1877 (*see page* 219).
SEPOY (Review Order), 1877 (*see page* 219).
SEPOY (Review Order), 1897 (*see page* 229).
SEPOY (Marching Order), 1897 (*see page* 229).

VIII. REGIMENTAL BADGES AND CRESTS (1850–1896) (*see page* 231).

IX. QUEEN'S AND REGIMENTAL COLOURS, 1892 (*see page* 242). *Frontispiece.*

(*The Plates will be found at end of book.*)

LIST OF BOOKS, MANUSCRIPTS, ETC. CONSULTED

Author's Name.	Title of Book.	Date.
Captain Innes-Monro	A Narrative of the Military Operations on the Coromandel Coast from 1780–1784 . .	1789
Colonel William Fullarton	A View of the English Interests in India, and an Account of the Military Operations in the Southern Part of the Peninsula during the Campaigns of 1782, 1783, and 1784	1791
Anonymous . .	Narrative of the Operations of the British Army in India from 21st April to 16th July 1791	
Major Dirom's .	Narrative of the Campaign in India which terminated the War with Tippoo Sultan in 1792	1793
Edw. Moor's . .	Narrative of the Operations of Captain Little's Detachment, and of the Mahratta Army, during the late Confederacy against Tippoo Sultan	1794
Captain Charles Gold	Oriental Drawings (1791–8)	1800
. . . . . .	Annual Asiatic Registers from 1799	
Lieutenant-Colonel V. Blacker's	Memoir of the Operations of the British Army in India during the Mahratta War of 1817–1819	1821
. . . . . .	The East India Military Calendar for 1823–1824	
. . . . . .	East Indian Registers from 1810	
. . . . . .	Asiatic Journal and Monthly Registers from 1805	
Mill's	History of British India	1826
Dodwell and Miles' .	List of Officers of the Indian Army, 1760–1837	
Wilk's	Historical Sketches of Southern India . .	Repr. 1869
Colonel James Welsh	Military Reminiscences	1830
R. Ackermann's .	Costumes of the Indian Army	
Bucle's	Memoir of the Services of the Bengal Artillery .	1854
. . . . . .	General and other Orders from 1800	
. . . . . .	Madras and Indian Army Regulations	

Author's Name.	Title of Book.	Date.
. . . . . .	Regimental Manuscript Records	
. . . . . .	The Madras Mail	
. . . . . .	Journal of the R.U.S. Institution	
Colonel Wilson's .	History of the Madras Army	1882
Lieut. A. G. Burn's .	Record of Services of 14th Madras Infantry	
Lieutenant Rainey's .	Report on the Chins, etc.	
. . . . . .	History of the Third Burmese War	
Captain H. Lawson's	Manuscript Notes	1894
Captain J. A. Loudon's	Journal	1896
Lieutenant A. T. Kirkwood	Letters	1896
. . . . . .	India Office Manuscript Records	
. . . . . .	Rangoon Gazette	

13TH MADRAS INFANTRY.

	13TH CARNATIC BATTALION . .	1776
Became	13TH MADRAS BATTALION . .	1784
,,	2ND BATTALION, 3RD REGIMENT .	1796
,,	13TH REGIMENT NATIVE INFANTRY .	1824
,,	13TH REGIMENT MADRAS INFANTRY .	1882

PART I

REGIMENTAL HISTORY

HISTORICAL RECORDS

OF THE

13TH MADRAS INFANTRY

I

REGIMENTAL HISTORY

IN the year 1640, when Fort St. George was built at Madras, the merchants employed armed retainers, known as "Topasses" and "Mistices" to the old writers, for the protection of their factories. The "Topasses," so called from their wearing hats instead of turbans, were native Christians who dressed like Europeans, and called themselves Portuguese. It was not until about a century later, namely, in 1746, that any attempt was made to raise and organize troops. England was then at war with France, and in this year Madras was besieged and capitulated to the French. The local British Government of the day having been compelled by this disaster to betake itself to its other settlement, Fort St. David, near Cuddalore on the Coromandel Coast, began to raise forces for the maintenance of its position against the French. These forces seem, however, to have possessed but little organization, and were known as "sepoys" and "peons." The former, though armed and equipped like Europeans, retained their own dress and customs. The "peons" were a body of irregular troops, whose strength depended on the circumstances of the time and the amount of funds

2

available for their up-keep. These men served their purpose, supplemented by the Topasses, Negroes and Arabs from Bombay, Rajpoots, Hindostanees, and any fighting men that could be obtained. At first these levies were composed entirely of such foreigners, and the following description is given of the system under which their services were utilised :—

> "The first sepoy levies had no discipline, nor any idea that discipline was required. They were armed with matchlocks, bows and arrows, spears, swords, bucklers, daggers, or any other weapons they could get. They consisted of bodies of various strength, each under the command of its own chief, who received from Government the pay of the whole body, and distributed it to the men, or was supposed to do so. Sometimes those chiefs were the owners of the arms carried by the men, and received from each man a rupee a month for the use of the weapons."

1758 It was not until twelve years later, namely, in August 1758, that any attempt at regular organization was made, when, urged on by the fall of Fort St. David, and by the absence of the greater portion of the troops, which had been sent on a sudden emergency to Bengal with Clive, and by the necessity of opposing, with a disciplined force, the French who were marching on Madras, companies of 100 men, each with a complement of native officers and non-commissioned officers, were 1759 rapidly organized, and the following year (September 1759), the siege of Madras by the French having been raised some months previously, these companies were formed into five battalions. Two of these battalions would appear, however, to have been formed the year before, for the returns of the Fort St. George garrison during the siege show that at that time two battalions, strength 2213 men, formed part of the troops, and lost 105 killed and 217 wounded. These battalions were composed of inhabitants of the Carnatic, and in this way the present Madras sepoy force came into existence.

1766 It is not until the year 1766 that a 13th Regiment was found in the Company's service. This battalion was raised for employment in the Northern Circars, and there is good cause for believing that, under Captain Craigneau, it formed part of Colonel Smith's force at the battle at Chengamah on the 2nd September 1767, when he defeated Hyder. This was the first general action in which Madras sepoys took part. 1767

This 13th Battalion, however, was but short-lived, and became the 3rd Circar Battalion on the 16th June 1769. It is not now represented in the Madras Army. The 16th Battalion then became the 13th Carnatic Battalion, which, on the reduction of the 3rd in 1770, was renumbered the 12th. 1769

From 1770 to 1776 there was no 13th in the Madras Army, but in the month of December of the latter year the strength of each battalion was reduced from 1000 men to 770, and the surplus was formed into six additional battalions. Of these, the 13th Carnatic Battalion was formed at Madras of drafts from the 4th, 7th, and 11th Battalions, and is now represented by the 13th Madras Infantry. On the formation of the battalion, Captain Hugh Robert Alcock was appointed Commandant, and Lieutenant Richard Anderson, Adjutant. The 13th Madras Infantry is still known in the Native Army as "Ulka's Pultun" (Alcock's Regiment).[1] 1776

The following is the quaint description of the drill directed to be carried out at the time of the formation of the regiment :—"A form of exercise the most simple and easy, as likewise evolutions such as will be best adapted to the nature of these troops and the intended use of them, will be pitched upon, and when chose, no other to be followed, unless such changes as the superior officers may think proper to order."

[1] "The Madras Infantry Regiment first raised in 1758 was called by the natives *Lall Pultun*, or the red battalion; Pultun being a native corruption of the word platoon, itself derived from the French *Peloton*, being a detachment of about 30 men."—*Carnac.*

1778 LIST OF OFFICERS OF THE 13TH CARNATIC BATTALION AT THE END OF 1778 OR BEGINNING OF 1779

(*From the Manuscript Records in the India Office.*)

Hugh Robert Alcock		Captain 4/5/1768	Commandant 12/1776
George Wahab .	Ensign 9/11/1770	Lieutenant 28/7/1775	Adjutant 1778
Caleb Pearson . .	Ensign 10/12/1770	Lieutenant 26/8/1776	
John Spencer . .	Ensign 18/8/1773		
George Aubery .	Ensign 9/9/1773		
Thomas Sampson .	Ensign 5/6/1775		
Andrew Boswell .	Ensign 20/9/1775		
Thomas Davies .	Ensign 21/8/1776		
Thomas Phillips .	Ensign 10/1/1778		
William Collins .	Ensign 11/5/1778		
Harcourt Powell .	Ensign 4/8/1778		
William Wemyss .	Ensign 11/8/1778		
Samuel Godfrey .	Ensign 25/8/1778		
William Fenwick .	Ensign 27/8/1778		

In June 1778, the 13th Carnatic Battalion, with seven others, were warned for service against the French settlement of Pondicherry, and on the 8th August, with the main army, encamped near that place. Batteries were opened on the 18th September, and after an obstinate defence, lasting one month, Pondicherry capitulated on the 17th October. 265 pieces of ordnance, 6000 muskets, and 100 rifles were captured. The loss on both sides was heavy, and that of the 13th was returned as 1 ensign, 1 subadar, and 6 sepoys killed; and 1 captain, 1 jemadar, 2 naicks, and 12 sepoys wounded. As the establishment of a Carnatic battalion included one captain only, it must be concluded that Captain Alcock was the captain wounded on this occasion. Immediately after Pondicherry capitulated, the demolition of the fortifications was begun, and completed about 1779 the end of the following year (1779), during which time the 13th Carnatic Battalion presumably took part in the work. If so, however, the battalion must have shortly afterwards moved into garrison at Tanjore, for, on the 1780 appearance of Hyder, on the 24th July 1780, with his army before Madras, Colonel Cosby was sent to the south to organize the famous

TRICHINOPOLY DETACHMENT,

to which the 13th Battalion, then stationed at Tanjore, contributed its two grenadier companies. This Detachment, according to Colonel Wilson, was organized to intercept any convoys which might come through the passes for Hyder's army. It was composed of two cavalry regiments, the grenadiers of the 6th, 9th, 13th, and 19th Battalions, and three battalion companies of the 19th Battalion. Failing to fall in with any of Hyder's convoys, Cosby made up his mind to attack the fort at Chittapet. He attacked on the night of the 6th September, and endeavoured to carry it by escalade, but the enemy being prepared for his reception, he was repulsed with loss. Cosby then marched on Chingleput, and there joined the main army under Sir Hector Munro, who was retreating on Madras after the destruction of Colonel Baillie's force by Hyder on the 10th September at Perambakum. The whole force retreated to Madras, and encamped at Marmelong on the 14th. The Trichinopoly Detachment, under Captain Bilcliffe (with the exception of the grenadiers of the 9th and 18th Battalions, who had rejoined their corps), occupied the posts at the Great and Little Mounts until the assembly of the army early in January 1781, under Sir Eyre Coote, when they marched on the 17th with the remainder of the army for the relief of Chingleput, Wandiwash, and Permacoil. After the relief of these garrisons, the army marched to Pondicherry, and remained in that neighbourhood for several months almost inactive, and much distressed for want of provisions. The Trichinopoly Detachment took part in the capture of the fort at Tiruwadi in April, under the command of Coote, and in the attack on the fortified pagoda at Chillumbrum, which had been strengthened by Hyder with the intention of using it as a depôt for provisions. The

1781

attack on this place was repulsed with considerable loss.

Sir Eyre Coote now determined to commence a serious attack upon the fortress, and started for Porto Novo to procure battering cannon, which had just been conveyed there by the British fleet. The camp was no sooner pitched at Porto Novo than it was instantly surrounded by the whole Mysore Army. Preparations were made on both sides for battle. The British General sent his heavy baggage on board the fleet, and at four o'clock on the 1st July the line was put in motion. "At the same time the enemy were seen arranging in battle array, forming a force computed at one hundred thousand men, including two troops of French hussars, and a battalion of European renegadoes."

Captain Innes-Monro describes the Battle of Porto Novo as follows:—

"When our army, which consisted of nearly 8000 effectives, had filed off from the right, and advanced about a mile into a large plain, it was formed into two lines. The first, under Sir Hector Munro, was composed of 3 regiments of Europeans, 6 battalions of sepoys, 1 European troop, and 2 regiments of black dragoons, with 30 pieces of cannon. The second, led by Major-General Stuart, was composed of 4 battalions of sepoys, and 26 field-pieces—1 battalion of sepoys, and 2 regiments of cavalry with 6 guns, were allotted for the baggage guard.

"The British troops were first menaced by a chain of Hyder's irregular horse, who appeared immediately in our front as a blind for their masked batteries; by whom innumerable quantities of rockets were thrown, and also some shot. This induced the British line to advance towards them without hesitation, exchanging a few random shots in their progress; but as we approached towards them they gradually retired, and at length suddenly disappeared behind the woods, at once unveiling to us six large masked batteries of 6 or 10 guns each; from which, however, no shot was yet fired. Upon this discovery the army made a halt until the ground should be closely reconnoitred, when it was judged inconsistent with prudence to advance any farther in front. Our line was therefore ordered to file off to the right, which the enemy no sooner perceived than all their batteries made a furious opening and galled us unmercifully. Having at length cleared the range of these batteries, we kept along a sandhill betwixt us and the enemy's line, which had by this time quitted its former station, and now moved at some distance in a parallel line with us. In the centre of this sandhill there appeared a narrow pass, through which our first line bent its course and formed upon a plain on the other side, but not without suffering severely from the Misorian line of infantry and guns drawn up directly in our front, behind a range of low sandhills and bushes, from whence they fired warmly into the pass. A bloody conflict took place between the enemy and our second line for the possession of the hill upon our left flank, which was at last gained by the point of the bayonet. By this time the first line had got completely formed upon the plain, and at about 10 o'clock a dreadful cannonade commenced on all quarters, and continued without intermission until three in the afternoon. The superior excellence of our artillery had visibly much slackened the enemy's fire and thrown their ranks into disorder. The enemy endeavoured to cover the retreat of their guns, but our guns being loaded with grape-shot, poured it in amongst the enemy's cavalry and infantry without mercy. The British continued to pursue them until we reached the summit of the sandhills, where, being stopped by a deep nullah from advancing any farther, a few more shots put an end to the carnage and fatigues of the day, with the loss on our side of 17 Europeans and 20 black officers, besides 50 Europeans and 500 sepoys, killed and wounded. Hyder and his army reached Chillumbrum before they once entertained a thought of rallying. As soon as the wounded could be collected together, and the dead interred, our army crossed the nullah, which was in many places so filled with slaughtered enemies that our soldiers could not avoid trampling upon their bodies as they passed. We lay that night on its opposite banks, and next morning encamped upon the road leading to Cuddalore, where a *feu de joy* was fired, and the General's thanks delivered to the army."

The Trichinopoly Detachment was in the first line at the Battle of Porto Novo, which was one of the decisive battles of India, according to Colonel Malleson; and the first, it is said, that caused Hyder Ali to despair of driving the British out of Southern India.

The following extracts are from Sir Eyre Coote's report of the battle:—

"The bravery of our troops at length carried the point, and the first line forced the enemy's infantry, artillery, and their cavalry to give way, obliging them to seek for safety by a retreat."

"The behaviour of the whole army on this most interesting day was uniformly steady, and worthy of the highest commendation."

"The spirited behaviour of our sepoy corps did them the greatest credit. No Europeans could be steadier; they were emulous of being foremost on every service it was necessary to undertake."

After this battle Coote joined Colonel Pearse's Bengal Detachment at Pulicat, and the whole returned to the Mount, and were formed into brigades on the 8th August 1781. The Trichinopoly Detachment, together with the 13th Regiment of Bengal Sepoys, and the 9th, 17th, and 18th Carnatic Battalions, formed the 3rd Brigade under the command of Colonel Pearse (Bengal).

The British Army now made every preparation for a march to the siege of Arcot, and the relief of Vellore, which was besieged by Hyder. For this purpose the troops, furnished with only eight days' rice, began their march thither upon the 16th August. On the 20th they tôok Fort Trippassoor, which at the commencement of hostilities had fallen into the enemy's hands. Hyder, whose army was then encamped at Congeveram, endeavoured to dispute the road with the British. As the army advanced, his irregulars took every opportunity of obstructing its march; and, on the 27th August, exactly upon the same ground where Colonel Baillie's army had been annihilated the year before, he appeared in full force, strongly posted behind the woods and village of Pollilloor. The enemy's position was first announced by some guns unexpectedly opening fire upon the advanced guards.

The following is a description of the Battle of Pollilloor:—

"The fire from the enemy was soon increased into a very hot cannonade, under which our line was irregularly formed upon broken grounds, everywhere intersected by deep nullahs. A front had no sooner been presented, though in a manner hasty and confused, to the guns of the enemy in one direction, than fresh batteries were opened from other quarters upon both flanks and rear, by which means the line became divided and detached. The fourth brigade, under Colonel Pierce, was formed in a position to oppose the fire of our left flank; at the same time the rest of the line, under Sir Hector Munro, changed front twice; first to the right and then to the left, under a very hot fire; while the second brigade, led by Colonel Edmonston, was ordered to attack the village of Polliloore, then opposite to our right flank. General Munro afterwards advanced with his division against the village, a measure which was attended with immediate success. The enemy being, by these manœuvres, at length dislodged from every strong position on their left, Colonel Owen received directions to move with four battalions from our right wing to assist those upon our left in throwing the enemy's right flank into confusion; which, with a few more evolutions of the first line, attended with an incredible share of fatigue, soon put the whole to flight, and finished the action at sunset.

"That night the British forces encamped upon the field of battle, which was the only pretension we had to term this a victory. A *feu de joie* was, however, fired by both armies, as neither was willing to yield to the other the laurel of the day."

In this battle the 3rd Brigade formed part of the

first line, and contributed in no small way to the success of the day.

The following extract is from Sir Eyre Coote's report of the fight :—

"The fire from Colonel Pearse's quarter was, during the general cannonade of the afternoon, of great importance; they had frequent opportunities of directing it towards the flag elephants, whose standards were seen over the rising ground where Hyder Ally himself was posted. The shot fell frequently amongst them."

Hyder's army retired in great confusion beyond Congeveram, leaving a flying camp of cavalry behind to watch Coote's movements, but, on the advance of Pearse's brigade the next morning, they "broke up and precipitately retired to their main army." According to Coote, Hyder's force numbered 150,000 men with 80 pieces of cannon, and their loss was nearly 2000; whilst of Coote's army of 11,000 Europeans and natives combined, the casualties were only—

Europeans killed	. 28	Natives killed	. 105
,, wounded	25	,, wounded	207
,, missing	. nil	,, missing	. 58

After two days' halt on the ground, the army retired to Trippassoor for fresh supplies. On the 19th September, the army made a movement towards Vellore *viâ* Pulicat and the Sholinghur Mountains, which protected the right flank and baggage.

"Hyder was now desirous of trying his utmost force at the Pass of Sholangur; where, having taken every necessary precaution, he again waited our approach in battle array. He seemed fully determined to make every opposition to the relief of Vellore, which was unremittingly invested by strong detachments of his forces. . . . Hyder's camp was pitched upon a very gentle declivity, with strong grounds and a tank of water in his front; his left flank being posted under the Sholangur Hills.

"The British Army having advanced within a few miles of its adversary, encamped with its right flank covered by the same hills, and in a parallel situation with the enemy's line; which was, however, completely separated from ours by a long range of low rocks, which ran down from the mountain of Sholangur betwixt the two armies.

"Early in the morning of the 27th of September, Sir Eyre Coote, with the second brigade and some cavalry, advanced to these rocks, and reconnoitred the enemy's camp, which he found still pitched, though their forces were forming in order of battle, and large bodies of horse seemed already advancing upon our left. Orders were then issued for the British Army to be brought up exactly in front.

"A brisk cannonade began on both sides, which was on our part very warmly kept up; while the second brigade, joined by the cavalry and the guns, crept round the tank unobserved, and fell suddenly with all their fire upon the camp and left flank of the enemy; the rest of the army at the same instant advancing, and supporting a warm discharge of shot and grape, which threw the infantry and cavalry of that wing into confusion. Hyder, perceiving this, tried a similar experiment upon the left of our line; in which attempt, however, they were so warmly received by the British artillery as to have their whole army thrown into general disorder, upon which Hyder thought it full time to draw off his guns. As a last effort, a thousand desperate horsemen charged the English line where four battalions of sepoys were stationed, but these sepoys reserved their fire until the assailing horse were within fifty yards of them, when they delivered a well-levelled volley with fatal execution. The horsemen, however, forced their way through the intervals; but so far were they from breaking the battalions, that the latter went quickly to the right about, and gave the enemy a second fire in the rear, which caused them to pay a bloody price for their temerity.

"By this time the enemy's line had become completely routed; the second brigade and British cavalry continuing the pursuit till sunset, which put an end to the action. It was computed that the enemy had upwards of two thousand men and horses killed and wounded on that day. Ours, not exceeding one hundred, officers and men included, was but a trifling loss."

It was at Sholinghur that the present 20th Madras Infantry, then the 21st Carnatic Battalion, and now one of the linked battalions of the 13th Madras Infantry, earned the unique distinction, still enjoyed by them, of carrying an extra colour, and of an extra jemadar to carry it. This reward was conferred upon them because the battalion, being charged by Hyder Ali's select corps of stable horse, as related above, received the assault with a "steady coolness which was highly conspicuous," and drove it off, capturing two standards.

As the troops had apparently been brigaded afresh, it is difficult to say what part the Trichinopoly Detachment took at the Battle of Sholinghur.

The army at this time was in great straits for supplies, and Sir Eyre Coote reported to Government that "never since he had been a soldier, which was then forty years, had he seen such distress in any army as then prevailed in his."

"After the Battle of Sholingur, Sir Eyre Coote determined to march through the Pollams, under the auspices of the rajah, in quest of a daily subsistence, and to trust to Providence for the relief and protection of Vellore. The British Army, with only one day's rice, began its march on the 1st of October through the Sholangur Pass; the second brigade being posted there as a guard; and, after a fatiguing march of two days amongst the hills, pitched its camp at Attamancherry. Here the army lived in plenty and tranquillity for the space of several days; but on the 12th of the month, intelligence was brought to the General, that about six thousand of the enemy's horse had entered the Pollams by a secret pass, and had begun to plunder the villages. Sir Eyre Coote instantly marched at the head of three battalions of sepoys, and all the cavalry; and completely surprised the whole of this party in their camp.

"As soon as a sufficient portion of rice could be collected for the relief of the immediate distress of Vellore, Colonel Owen, with one hundred European grenadiers, five battalions of sepoys, a regiment of cavalry and ten guns, was ordered to advance before the army, and take post at the Pass of Veracundaloore in order to protect the Polligars and others who endeavoured to supply the garrison of Vellore with grain. Hyder unexpectedly reached Owen's camp on the 23rd October with the whole of the Mysore Army. It was with difficulty that our troops could snatch up their arms and form the line, before they were beset upon all quarters. The tents and baggage were burnt, and the chief object of the contest was first to gain possession of the pass betwixt Colonel Owen and our main army. Detachments were made from both parties with the greatest celerity for this purpose, but fortunately effected by ours, which soon brought on a close and general engagement. Hyder, however, was compelled to draw off his troops, much to his disgust, as he made certain of being able to cut off this detachment. Colonel Owen soon afterwards joined the main army, which encamped that evening at the village of Madowaddy."

Upon the 26th October the General moved his camp to Paliput, whilst a detachment was sent back to Trippassoor with the sick and wounded of the army. A large quantity of rice was discovered at Paliput, which at last enabled the British, on the 3rd November, to relieve the inconceivable distress that Vellore had experienced for a considerable time. During this march the enemy gave very little trouble. The main army remained at Vellore for two or three days, and, after being reinforced by Colonel Laing, with 100 European grenadiers, proceeded to the attack of Chilloor, which made a gallant defence for some days, but capitulated upon a breach being effected in the rampart. On the 2nd December the army received orders to break up their camp on the Cocolore Plain, and march into cantonments in the environs of Madras.

1782 The object of the first military operations of 1782 was the relief of Vellore, which had been constantly blockaded by the enemy, and had almost expended the supply of three months' rice which the army had with great difficulty thrown into it at the close of the preceding campaign. Vellore was now the only key that remained in the British possession to the passes leading to the enemy's country.

Sir Eyre Coote marched the army to Vellore *viâ* the Sholinghur Road, in order to have his right flank and the convoy covered by the Pollam Hills. Hyder disputed the way, but the garrison was relieved on the 11th January, and supplied with six months' rice. Coote lost 3 officers and 70 men killed and wounded during the skirmishes on the march. Exasperated by his failure, Hyder determined to revenge himself by setting a trap for the British Army on their return to Madras. He accordingly gave orders for some sluices to be opened in the bank of a tank below which the troops must return, thus inundating the country and rendering it almost impassable for infantry. On the 13th the army departed from Vellore, and, without the least suspicion, entered the slough, which it was permitted quietly to pass, until the European brigade had got quite entangled in the mud. Upon this, upwards of 50 guns furiously opened upon the British, who immediately pushed forward and formed upon the other side. Hyder's design being frustrated, he retreated to Arcot, leaving the British to continue the march unmolested to Poonamalee, which was reached on the 20th. Six officers and about 30 Europeans and 100 sepoys were killed and wounded in this expedition.

The Trichinopoly Detachment during 1782 formed part of the main army. On the 2nd April 1782 the French, having been reinforced to the extent of some

3000 men, appeared before Cuddalore, which was garrisoned by the 12th Carnatic Battalion under Captain James Hughes, and, having offered favourable terms, the place was given up to them on the 3rd or 4th without a shot being fired. Shortly afterwards, the hill fort at Permacoil, garrisoned by one company of the 16th Battalion, fell into their hands. Coote does not appear to have shown much enterprise on this occasion, for although he marched against Hyder and his allies the French, he did not venture to attack them, but, instead, threatened Hyder's depôt at Arnee. Hyder, alarmed for the safety of his depôt, detached Tippoo, who had joined him after his success against Braithwaite in February, with orders to reinforce the garrison, whilst he himself followed Coote and brought on an engagement near Arnee, under cover of which, Tippoo was enabled to effect his purpose. The battle at Arnee, although Hyder is stated to have been beaten there, does not appear to have secured any advantage to Coote, for on the 18th June the army returned to Madras. Sir Eyre Coote, however, was at this time suffering from ill health, and on the 28th September he embarked for Bengal in the frigate *Medea* for change of air. This was practically the last of Coote's soldiering, for, having returned to Madras on the 24th April, he died three days afterwards, much to the grief of the army.

Sir Hector Munro having previously resigned, Major-General James Stuart succeeded to the command of the army.

On the 7th December 1782, Hyder died at Chittoor of carbuncle, and his body was removed by Tippoo to Seringapatam.

1783

On the 1st January 1783 the army at the Mount was formed into two lines, and the Trichinopoly Detachment, with which the two grenadier companies of the 13th Battalion were still serving, formed part of the

3rd Brigade of the 1st line. The army moved south on the 4th February, and on the 13th sighted the united forces of Tippoo and the French near Neddingul. On the approach of the British they retired, and Tippoo, hearing of the capture of Bednore by the Bombay troops, shortly afterwards returned to his own dominions in the west. General Stuart, having first demolished the works at Wandiwash and Carangooly, and again relieved the garrison at Vellore on the 21st April, commenced his march against the French at Cuddalore, but so dilatory were his movements, that he did not reach that place, a distance of only 126 miles, until the 7th June. Arrived before Cuddalore, the army was drawn up in the same order as it had been formed at the Mount in January. This order appears, however, to have been changed before making the assault on the French outworks on the morning of the 13th, for on that occasion the Trichinopoly Detachment formed part of the centre attack under Major-General Bruce and Colonel Gordon. The attack commenced before daylight, and firing was kept up until 4 or 5 o'clock in the afternoon. During the succeeding night the enemy abandoned all their outworks, and retired into the fort, taking their heavy guns with them, but leaving 15 or 16 field-pieces in the possession of the British. The behaviour of the Sepoy Battalions was highly praised. The loss of the British was 588 Europeans and 347 natives killed and wounded, of which the Trichinopoly Detachment lost 2 rank and file killed, and 1 subadar, 1 jemadar, and 24 rank and file wounded. The French were, however, reinforced by some 2400 men, and the position of the British became critical. Intelligence was shortly afterwards received of the "conclusion of peace in Europe, in consequence of which hostilities ceased on the 2nd July."

The two grenadier companies joined regimental headquarters at Dindigul on 23rd September 1783.

The following extracts from Captain Innes-Monro's letters, written about 1780, concerning the officers and men, and duties of native corps in the field, may not be uninteresting in these days :—

"All the Company's European officers are promoted by regular rotation ; which, with the frequent opportunities they have of seeing service, gives them a vast fund of professional knowledge. They are fortunate who arrive at a company after twelve or fourteen years' service, by which time their exemplary and assiduous attention to duty and discipline renders them fit to be entrusted with the most important command. Those who have by this means distinguished themselves in the service seldom fail of being recompensed in their advanced years with some handsome appointment, by which they can with care live comfortably all the rest of their lives. But, to speak truly, this independence is literally earned by the sweat of their brows."

"These black corps have attached to them a full complement of native as well as European officers. The former rise according to their merit from private sepoys ; and before the most of them arrive at the rank of subidars or captains (for higher they do not go), they become quite bald and grey in the service ; and their hoary beards and whiskers cut a most venerable appearance at the head of a regiment. Their rank gives them no authority except over their own countrymen ; for a European sergeant would command any of the native officers upon duty. . . . All words of command are given in English ; and each battalion has a good corps of drums and fifes."

"Every sepoy in the army carries with him to camp his whole family, be they ever so numerous, who live upon his pay and allowances of rice from the Company. This practice, when properly considered, is really justifiable in them, for an Asiatic must have his wife, whatever may be his circumstances ; nor is it customary upon any occasion for man and wife to be separated. The wife shares the hardships of war with her husband in the most cheerful manner, let them be ever so perilous, and follows him wheresoever he goes. Besides, a sepoy's station in life is reckoned so far respectable and elevated above the common rank, that he is looked up to for support and protection by all his needy relations, which he generally affords them to the utmost of his power, sharing his all with these dependants, who have no other home but his barrack or tent."

"All the duties of advanced guards and pickets, escorts, and other sources of fatigue, are performed by the sepoys, who are surprisingly active upon their posts, and pay the most implicit attention and obedience to orders, which are given in English, but afterwards explained to them in their own language. The Europeans do no other duty in camp than mounting their own quarter guards, excepting when we are close upon the enemy, lest the heat should throw too many of them into hospital."

"The cavalry and sepoys are always put upon the rear-guard, commanded by a field officer ; and it is sometimes five or six hours, after the rest of the line have been refreshed and snug in their tents, before the rear-guard can drive all the baggage and followers of the army into camp, unless upon a day on which the enemy have been numerous and daring in the rear."

Regimental Headquarters.

It is now necessary to return to the headquarters and eight battalion companies of the 13th Carnatic Battalion, which had been left in garrison at Tanjore.

1780 In November 1780, Colonel Braithwaite assumed command of the troops in Tanjore, and for nearly four years afterwards there was constant fighting in the

1781 Southern Provinces, with varied results. Early in 1781 Hyder took possession of the whole district of Tanjore, with the exception of the capital, placed garrisons in most of the forts and defensible pagodas, and instigated the Polygars of the neighbourhood, as well as those of Madura and Tinnevelly, to rise in rebellion. On the 1st of August Colonel Braithwaite attempted to storm the fortified pagoda at Tricatapully, but was repulsed. On the 3rd of the same month he was again repulsed at the fort of Puttoocottah, and, having been wounded during the assault, he made over command to Colonel Nixon, who up to that time had been commanding the troops in Trichinopoly. About the 8th September 1781, Colonel Nixon attacked and took the fort at Manargudi, about 24 miles south-east of Tanjore. His force consisted of 116 artillerymen, European and native, 84 topasses, 166 native cavalry, a few Mogul horse, the 6th, 10th, and 13th Battalions, and a Tanjore local corps, amounting in all to about 3300 men.

On the 16th of the same month he attacked the fort at Mahadavypatam, which he described as "very strong and well garrisoned by resolute men." The place was taken after an obstinate resistance, and not without considerable loss, namely, 3 European officers, 3 sergeants, and upwards of 300 sepoys killed and wounded.

Colonel Braithwaite resumed command immediately after the capture of Mahadavypatam, and on the 30th September he defeated the enemy at Alangudi, about ten miles south of Combaconum, after a severe struggle. He described the place as standing like an island in the midst of paddy fields, and protected by deep water-courses. Two separate attacks were made, one under Colonel Nixon with the handful of Europeans, supported by the 6th and 13th Battalions. The other was composed of the 10th Battalion. The village was resolutely defended, but both attacks proved successful, and the enemy were driven out with the loss of one gun and a few tumbrils. Colonel Braithwaite estimated their strength at 5000 men and 6 guns. His own force, officers included, amounted to 2461 men and 8 guns. Of the enemy, 25 horses and 98 men were killed, and 70 horses and 217 men wounded. Two of their battalions were commanded by French officers, both of whom were taken prisoners; and the Colonel mentioned that a Captain Mills, who commanded two battalions, came over during the action. The British loss was comparatively small, namely, 19 men and 8 horses killed; 2 officers, 4 Europeans, and 64 sepoys wounded. Colonel Braithwaite spoke in high terms of the conduct of the troops, more especially of that of the cavalry under Lieutenant Sampson.

On the 21st October 1781, Nagore, about four miles north of Negapatam (both of which places were held by the Dutch), was, on the approach of Colonel Nixon's force, evacuated. The garrison was pursued by the cavalry under Lieutenant Sampson, and lost 200 men, 4 standards, 4 guns, and 2 tumbrils.

The same day Major-General Sir Hector Munro landed with 300 marines and 500 seamen, and assumed command of the Southern Army. Negapatam, the principal settlement of the Dutch in Southern India, was the next place to fall into the hands of the British. On the 29th October 1781, the outworks in front of Negapatam were assaulted and carried, and the following is an extract from Sir Hector Munro's report of the operations :—

"I have the honour to acquaint your Lordship, that last night we attacked the enemy's advanced posts, consisting of five redoubts with a line of communication to each, and carried them, with very trivial loss on our side. The enemy must have suffered considerably, but what their loss may be, I neither know nor care. Our killed and wounded amount only to 11 Europeans and 20 sepoys. We have taken 1 officer, 2 cadets, 13 Europeans, 79 sepoys, and 19 Malays. The enemy left behind them, in their redoubts, 19 pieces of cannon, with a quantity of stores. In this attack the officers displayed equal bravery and good conduct, and the men behaved with great resolution. There were, indeed, three attacks, two real and one false. The two former were conducted by Colonel Nixon and Captain Scott."

The following is Wilson's account of the capture of Negapatam :—

"Batteries having been constructed, the garrison was summoned to surrender, but the Governor refused, and two determined sallies were made, both of which were repulsed by the seamen and marines in the trenches. The batteries were then opened, and one of the bastions having been quickly destroyed, the place capitulated on the 11th November, and was taken possession of on the succeeding day. The property found in the citadel, consisting of ordnance, military stores, grain, and merchandise, was very considerable. The following are some of the principal items, viz. 188 serviceable guns, 8 brass mortars, 277 barrels of gunpowder, a large quantity of shot, shell, and hand grenades, 3241 serviceable muskets, 1514 swords, and a number of cartridge pouches and other accoutrements. A quantity of naval stores, such as sail cloth, anchors, tar, pitch, and leather ; also engineer's stores and tools. Amongst the various kinds of merchandise were 1410 bales of cotton cloth of different descriptions, 115,686 lbs. of cloves, sugar candy 84,456 lbs., sugar 173,011 lbs., Japan copper in bars 81,176 lbs., 537 bars of silver, and gold in bullion, and coins to the value of rupees 39,664."

From which goodly list the victorious troops were no doubt well remunerated. The native troops engaged were detachments of the 6th, 10th, and 13th Battalions, the Tanjore Light Infantry, detachments from other local corps then being raised at and near Tanjore, and a few troops of cavalry.

In December 1781, volunteers were called for from the 9th, 10th, 13th, and 23rd Battalions to form a battalion to accompany the expedition to Ceylon, which sailed from Negapatam on the 2nd January 1782, for service against the Dutch. All who volunteered for this

1782

service were either pariahs or men of the lowest castes. Trincomalee was captured on the night of the 5th January, and Fort Osnaburg on the 11th, but was surrendered to the French the following August.

After the fall of Negapatam, the enemy quitted the district of Tanjore, but in the month of February 1782, a terrible disaster overtook the force under Colonel Braithwaite, who was in camp near the village of Annagudi, six miles north-east of Combaconum. He had with him some 1700 men, out of which three officers, Lieutenant Lind, Ensigns Gahagan and Haywood, and 300 men, belonged to the 13th Battalion. Taken completely by surprise by Lally and Tippoo, though, according to Wilks, he had ample warning, which he altogether disregarded, after twenty-six hours of desperate fighting he suffered a crushing defeat, and his force surrendered at discretion. The following account is taken from Wilson's *History*, and is an extract from a report written by Lieutenant Charles Salmon of the Engineers, who had been appointed aide-de-camp to Colonel Braithwaite, and was on his way to join him in camp:—

"Before you receive this you will no doubt have heard of the unhappy fate of our southern army, which surrendered at discretion on the 18th at twelve o'clock, at Annagudi, a village about four miles from Pantanellore. Tippoo Saib and Lally, with 600 horse, 12,000 infantry, and 20 guns, came upon them before they had timely notice to retreat. The Colonel attempted to retire in the night (having had a very hard day of it, and although surrounded, had kept his ground), but it was then too late, being surrounded and closely watched by the enemy. He marched about ten with two battalions of sepoys, leaving the cavalry and 13th Battalion on the ground, who were ordered to make a show of entrenching themselves by digging with mamoties, etc., and in two hours after to follow him as expeditiously as possible, leaving their tents standing. I am of opinion his intention was to retreat to Negapatam as he advanced towards Mayaveram, but he soon found that the enemy had got between him and the 13th Battalion. He therefore returned in order to join them, which he effected with great difficulty, being obliged to fight all the way back. The enemy rocketed him very much, which not only created great confusion, but did great execution. It was daybreak before the Colonel joined, when they had guns opened from every quarter on them. They sustained this heavy cannonade and were likewise engaged with the enemy's infantry till eleven o'clock, when the Colonel, finding that his destruction would be inevitable if he remained longer on that ground, marched off, and pushed for the pagoda of Manargudi (about a mile from camp), and he had nearly gained it, when one of the battalions (some say the Tanjore grenadiers), seeing a large body of the enemy advancing to charge them, made a run for the pagoda, leaving their guns behind them. This occasioned almost a general confusion, when the enemy cut in amongst them, and did great execution. At this time, Colonel Braithwaite received a wound from a horseman across his back. The 13th Battalion, that had the rear, behaved extremely well, and made their way good to the pagoda. The sepoys, who had now been two days without provisions, and fighting almost the whole time against such a superior force, were so disheartened that they called out for "cowle" (*i.e.* terms of surrender). There was a consultation of the officers, the result of which was, that they offered to surrender at discretion. There are several black officers, and also 500 sepoys, who escaped from the enemy, that are come in here. All the officers except Lind are wounded. I wrote to Captain Alcock and Mr. Sullivan all the particulars I could learn. Captain A. wrote me, that if I heard nothing from the Colonel, to retreat immediately, which I did, and had the good luck to get into the fort, bringing in the two companies from the fortified pagoda, and three which Ensign Salmon had at Triviar."

Captain Alcock, mentioned in the above, was the Commandant of the 13th Battalion, who had been detained in Tanjore in command of the garrison. Lind, too, who is mentioned by Lieutenant Salmon as the only officer not wounded, belonged to the 13th Battalion. The officers known to have been present in this action numbered twenty-one, including Colonel Braithwaite, and of these, Ensign Stuart of the 10th Battalion appears to have been the only one killed, so that the other nineteen must have all been wounded. These officers were all sent to Seringapatam, and remained there until the peace in 1784. With the exception of Ensigns Gahagan and Haywood wounded, it is not known what other casualties there were in the 13th Battalion. Lieutenant Henry Webber (who succeeded to the command of the regiment in 1810) escaped being present at this disaster, by being compelled from severe indisposition to go into sick quarters.

Throughout the whole of these campaigns the sepoys were half starved and in long arrears of pay.

During the year 1782, the 13th Battalion, which had apparently lost a considerable number of men in Braithwaite's disaster at Annagudi, was brought up to strength by drafts from some of the local corps raised during the war (probably the Tanjore Battalion), and orders were given for the formation of independent companies in different stations, from which to draft men to fill up vacancies, as they occurred in the regular army.

Colonel Braithwaite having been taken prisoner at Annagudi, the command of the troops in the south devolved again on Colonel Nixon, but, on the 23rd September 1782, he was superseded by Colonel Ross Lang, who was appointed to the command of Tanjore and all the troops south of the Coleroon. Having demolished the fortifications at Negapatam, he proceeded to Trichinopoly to prepare operations against Tippoo in Dindigul and Coimbatore. Moving out of Trichinopoly about the middle of March 1783, batteries were opened on the pettah at Caroor on the night of the 21st, and, a breach having been made, the place was assaulted and captured the next day, the enemy retiring into the fort, which was captured on the 2nd April. During the siege, Lang lost 158 killed and wounded. The fort at Avaracoorchy was captured on the 10th April 1783, and 400 of the enemy were killed. The fort at Dindigul surrendered without opposition on the 4th May.

1783

Shortly after this Colonel Lang was superseded by Colonel Fullarton, who at once augmented his force by battalions from Tanjore, Trichinopoly, and Tinnevelly. On the 25th May 1783 he marched from Dindigul, and encamped before Darapooram on the afternoon of the 30th, the last march having taken twelve hours. At three o'clock on the following morning he proceeded with some Europeans and two battalions of sepoys to take possession of a very strong post on the western side of the river, within three hundred yards of the fort of Darapooram. The men were under cover before daylight; a small mortar battery was opened at eight in the morning, a three-gun battery before three in the afternoon, and a breach was effected about six o'clock. Parties were sent round the fort and pettah to prevent the escape of the enemy, who did not venture to stand a storm. The fort was taken possession of on the 2nd June, and afforded ample supplies of grain and cattle. It was found to be capable of considerable defence, but Fullarton, not being able to afford sufficient men to garrison it, was obliged to destroy the fortifications.

Colonel Fullarton was about to move on Coimbatore, when orders arrived for him to join the main army at Cuddalore. Marching *viâ* Avaracoorchy and Caroor (where mines were constructed and the works blown up), he arrived at Trichinopoly in June. On his arrival the troops were supplied with grain, the gun-carriages repaired, cattle collected for the army at Cuddalore, and boats provided for crossing the Cavery and Coleroon, which, with two intermediate streams, were then unfordable. On Fullarton's arrival at Munsurpet on the northern bank of the Coleroon, he received further instructions from General Stuart to hasten his march to Cuddalore. Upon his arrival within three forced marches of that place, he received information of the cessation of hostilities between the French and the British. His help being no longer required in that direction, he turned south in July and returned to Munsurpet, intending to move next day by Tarriore to reinstate the Rheddey in his paternal inheritance, and to advance against the forts of Settimungulum, Namkul, and Sankerrydurgum, on the north of the Coleroon. His designs were, however, frustrated, as news was brought of an armistice with Tippoo Sultan.

Fullarton then returned to Trichinopoly, and decided to proceed to Dindigul for further orders. On the 25th and 27th July 1783, the army marched in divisions from Trichinopoly by the route of Touracourchy, as there was not sufficient water by the shorter road of Manapar. At Touracourchy, he received a requisition to move the army to Mellore in the Madura district, in order to awe the rebellious Colleries of that district, who, since the commencement of the war, had thrown off all appearance of allegiance. He arrived at Mellore on the 2nd August, and left a strong body encamped there. On the 4th August he marched to Shevigunga with a large escort, whither two battalions had previously been detached, in order to enforce the payment of the tribute due in that country. After an easterly march of about twenty miles, he found that the Rajah and his ministers had fled to the woods of Callacoil, and had collected a force of about 10,000 men. He could not prevail upon them to return, but in the course of four days he brought them to an arrangement by which they paid 40,0co rupees, and gave security for the remaining 50,000 rupees demanded.

At Shevagunga, Fullarton received instructions to move the army into the Tinnevelly country in order to repress the insolence of the Polygars, who had rebelled at the commencement of the war, and who had been committing daily ravages from Madura to Cape Comorin. They had subdued several forts, had occupied districts belonging to the Circar, and had paid no tribute. Fullarton determined to strike an unexpected blow against Catabominaigue, the Polygar of Panjalum-coorchy. The usual route to Tinnevelly passes by Madura; and the Polygars, hearing of the movement of the army towards Shevagunga, expected them in that direction. To favour this opinion, provisions for the army were ordered to be prepared at Madura; the real intention was perfectly concealed; and the force moved off from Shevagunga, on the evening of the 8th

August, to Tripechetty, a place twenty miles distant on the southern border of the Mellore country. Fullarton there joined the remainder of the army, and, leaving the 7th Battalion and some irregulars under Captain Oliver to restrain the Colleries at Mellore and to collect tribute, proceeded next morning by Pallamerry, Pandalgoody, and Naiglapour to the fort of Panjalumcoorchy, which was reached at 2 p.m. on the 12th August. The length of the march from Mellore was eighty miles, and from Sheviagunga one hundred miles, and this distance had thus been covered in four days, the thermometer being frequently above 110 degrees Fahrenheit during the marches. The quickness of these marches had not given Catabominaigue time to return from the siege of Chocumpetty; but in his fort were found about 2000 armed men, who shut the gates and manned the walls. As soon as the line approached the fort, a flag was sent, desiring the head men to open their gates and hold a conference, but they refused. As Catabominaigue was very soon expected with 8000 men, it was resolved to attack them immediately. The 18-pounders were therefore halted in rear of an embankment; a hasty battery for four guns was constructed at 4 p.m.; and in three hours the British were ready to open on the bastion. Fire was opened on the bastion, but, owing to its thickness, it was decided to breach the adjoining curtain, and to render the defences of the bastion untenable by the besieged. The enemy "kept up a constant and well-directed fire, and, notwithstanding the utmost efforts of the besiegers, it was dark before a practicable breach was effected. The attack was therefore deferred until the moon should rise. The storming party consisted of two companies of Europeans, supported by the 13th and 24th Carnatic Battalions, and continued in the rear of the battery; the cavalry, the first, and light infantry battalions were posted at right angles with the other three salient angles of the fort, with detachments fronting each gateway, in order to prevent the besieged from receiving supplies or making their escape, while the other troops remained to defend the camp, which was within random shot." A strong hedge fronting the breach, and surrounding the whole fort (the usual Polygar system of defence), was removed; and about ten at night, with the advantage of bright moonshine, the storm commenced. The result was thus described by Colonel Fullarton :—

"Our troops, after they gained the summit of the breach, found no sufficient space to lodge themselves, and the interior wall having no slope or talus, they could not push forward from the summit as they advanced. The defenders were numerous, and opposed us so vigorously with pikes and musketry, that we were obliged at last to retire, and reached the battery with considerable slaughter on both sides. Immediate measures were taken to renew the charge, but the Polygars, disheartened with their loss, abandoned the place, and sallied forth at the eastern gate. The corps posted round the works were so exhausted by the preceding marches, that many of the fugitives effected their escape; the rest were taken prisoners. The breach was covered with dead bodies, and the place contained a large assortment of guns, powder, shot, arms, and other military stores, which were of course applied to the public service; forty thousand pagodas were also found, and immediately distributed to the troops."

Very acceptable must this batta have been, as the pay of the native troops and their European officers was at this time twelve months in arrears, and a ration of rice to each sepoy was his sole means of subsistence; the very condiments necessary for rendering it fit for food being procured on credit from the native merchants of the camp bazaars.

Having left five companies of the 25th Battalion to garrison Panjalumcoorchy, Fullarton proceeded to Palamcottah, and from thence moved with the army by Shankanakoil to Shevagherry. The chief of the latter place was the most powerful of all the southern Polygars, and had remained hostile to the English during the whole of the war, and committed many outrages that called for punishment; amongst others, that of the murder of Lieutenant Graham Campbell, who had fallen into his hands. On the approach of the army before the town of Shevagherry, the chief evacuated it, and retired to the thickets, nearly four miles deep, in front of his "comby," or mountainous stronghold, which is close to the Travancore boundary. The chief manned the whole extent of a strong embankment that separated the wood and open country. He was joined by Catabominaigue with other Polygars, and his force, thus augmented, numbered between eight and nine thousand men. The "comby" had never been attacked before, and was thought to be impregnable. It was situated in the recess of an amphitheatre of very high rocks and mountains, and defended in front by a very strong work, mounting eight guns, the thick jungle being intersected with barriers and ravines, and the whole being enclosed from the open country by the high embankment, which the chief had occupied. The Shevagherry chief and his associates were informed that the place would be immediately attacked unless the head Polygars of Tinnevelly at once gave in their submission, liquidated all arrears of tribute, and paid compensation for all depredations committed since the commencement of the war. They replied that they wished to confer with Fullarton, but refused to visit him in camp. Fullarton agreed to meet them alone at their own barrier. Before they finished their explanations, it was dark, and a musket inadvertently fired in the rear alarmed the British advanced sentry, who thought it was aimed at him. Fullarton then withdrew.

The following is his account of the subsequent operations :—

"We refrained from hostility next day, but finding that they trifled with proposals, the line was ordered under arms on the morning following, and we made the distribution of attack. It proved as desperate as any contest in that species of Indian warfare ; not only from the numbers and obstinacy of the Polygars, but from the peculiar circumstances which had acquired for this place the reputation of impregnability. The attack commenced by the Europeans and four battalions of sepoys moving against the embankment which covers the wood. The Polygars in full force opposed us, but our troops remained with their firelocks shouldered under a heavy fire until they approached the embankment ; there they gave a general discharge and rushed upon the enemy. By the vigour of this advance we got possession of the summit. The Polygars took post on the verge of the adjoining wood, and disputed every step with great loss on both sides, After reconnoitring, we found that the comby could not be approached in front—we proceeded, therefore, to cut a road through the impenetrable thickets for three miles, to the base of the hill that bounds the comby on the west. The pioneers laboured with indefatigable industry, supported by the Europeans and 3rd and 24th Carnatic Battalions. The field-pieces were advanced as fast as the road was cleared. These were strengthened by troops in their rear, forming a communication with those in front ; for this purpose two other battalions were posted within the wood, and as soon as the embankment was gained, the camp moved near it, and the force concentrated. We continued to cut our way under an unabating fire from 8000 Polygars, who constantly pressed upon our advanced party, rushed upon the line of attack, piked the bullocks that were dragging the guns, and killed many of our people. But those attempts were repulsed by perseverance, and before sunset we had opened a passage entirely to the mountain. It is extremely high, rocky, and in many places almost perpendicular. Having resolved to attack from this unexpected quarter, the troops undertook the service, and attained the summit. The Polygar parties posted to guard that eminence being routed after much firing, on all hands, we descended on the other side, and flanked the comby. The enemy seeing us masters of the mountain, retreated under cover of the night by paths inaccessible to regular troops, and we took possession of this wonderful recess."

Eight guns, three elephants, and a large quantity of grain were captured. In his despatch to the Governor of Fort St. George, Fullarton reports as follows :—

"I cannot say too much in commendation of the officers and men in this business ; and am happy to add that, if we consider the strength of the place, our loss is extremely inconsiderable, as will appear from the list of the killed and wounded."

Leaving the 3rd and 9th Battalions to secure the magazines, Colonel Fullarton moved the army to Shevlepatore, within four marches of Madura, in order to awe the northern Polygars of Tinnevelly. Having there received the submission of the Polygars of that part of the country, he marched to Dindigul, where he arrived about the 23rd September. Here he was joined by Colonel Stuart's detachment, which included the two grenadier companies of the 13th Battalion, which had been detached to form part of the Trichinopoly Detachment since August 1780. The whole of the battalion was now present in Dindigul.

On the 25th September 1783, Colonel Fullarton, who had now between 13,000 and 14,000 men under his command, introduced several important changes in the method of encampment and order of battle. He told the troops off into five brigades, and formed the five brigades into two lines. The 13th Battalion (strength 698 fighting men), with the 6th, 8th, and 22nd Battalions, formed the 4th Brigade, under Lieutenant-Colonel Bruce, and were placed in the second line, which was commanded by Colonel Forbes. The light infantry of the 4th Brigade was placed on the left of the line, under the command of Captain M'Leod, and the grenadiers on the right of the line, under Captain Maitland. Two guns were posted on the flanks of the line, and between each battalion.

The native troops at this time obtained all their supplies from such magazines of the enemy as they were enabled to reduce, and up to the present it had been the custom of Indian commanders to levy duties on all articles bought or sold in the camp bazaars. This "odious tax upon the soldier for the benefit of his superior" was now abolished by Colonel Fullarton. The method and order of marching was also changed by him.

"The practice on the Coast has been to form the sepoy corps three deep, and the Europeans two deep, and then to move by files with a strong advance guard, and a still stronger rear guard, in order to cover the carts and other wheeled conveyances that follow the line; the baggage is then disposed of on the right or left flank, according to the nature of the ground over which the army is to pass, and covered by a strong force, to repel the rapid charges of the enemy's cavalry. It is asserted that many benefits attend this mode in India; and if the line is attacked on either flank, it is enabled to form with much celerity by a simple movement of conversion; and that if a charge be made in front or rear, the corps have only to advance, or countermarch, and form a front to the attack. But a large army marching by files is many miles in length, consequently there is little communication between distant parts of the line; neither can a commander observe the whole extent, or know the state of different divisions. If, in marching by files, a movement to the front or rear is necessary in line of battle or by corps, much time is lost in the manœuvre, and in the precautions requisite, in the face of an enemy. The Carnatic army, adhering to this principle, have frequently been cannonaded many hours before they could form the line for action.

"To remedy these evils, Colonel Fullarton proposed to form the army into five divisions, and to dispose them in shape of what in another science is called a quincunx. The European brigade, being usually placed in the centre of the line, should form the centre division of the quincunx, with a sepoy brigade in front, another in the rear, and one in each flank ; the battering train and baggage to move under cover of the division least likely to be charged ; and the brigades to move, not by files, but in columns, and at such distances, that whenever it may be necessary to form the line to the front, flank, or rear, the centre brigade, and that which is to become the right and left wings, may occupy the whole intermediate space. Thus, if the line be attacked in front, the centre brigade and the two flank brigades immediately form a line to the front, and the advance and rear brigades take their stations either as a second line, as a baggage guard and a reserve, or to extend the main line. If the line be attacked, or if it be meant to attack on the right or left flanks, the centre brigades, with the front and rear brigades, face to the right and left, and form the line ; while the two brigades that were the flank divisions on the line of march are posted as circumstances may require. Thus, in every possible point of attack, the line is quickly formed, the baggage is protected, and the army prepared for action."

The army having been directed to remain on the frontiers, ready to act offensively in case of an infraction on the part of Tippoo Sultan, moved for this purpose to reoccupy Darapooram. Here guns, shot, and stores were collected, and all preparations made for a movement on Seringapatam.

On the 16th October the army marched towards Pulney, about thirty miles south of Darapooram, in order to put the renter of Dindigul's family in possession of their inheritance. Here Fullarton received news of a recommencement of hostility on the part of Tippoo Sultan against Mangalore, and at once made preparations to march to the relief of that place, and from thence to Seringapatam, Tippoo's capital. He determined to proceed by way of Palghautcherry, and to garrison that place as a magazine for stores, or a line of retreat if necessary. Accordingly the army advanced from Pulney on the 22nd October, reduced the forts at Cumalum, Chucklegherry, and Annamully, and passed through a rich country abounding with dry grain, cattle, wood, and rice fields. During the whole march through this part of the country, the flank brigade, under Captain Maitland, moved constantly in front, occupied positions and secured provisions for the army. Colonel Fullarton thus describes this march :—

"From Annamully our progress became truly laborious, being obliged to force our way through a forest twenty miles in depth, extending thirty miles across the Pass of Palagat. Our object was to reach Colingoody, a post on the western side of the forest within fifteen miles of Palghautcherry. The frequent ravines required to be filled up before it was possible to drag the guns across them, innumerable large trees, which obstructed the passage, required to be cut down and drawn out of the intended track, and then the whole road was to be formed before the carriages could pass. The brigades were distributed to succeed each other at intervals, preceded by pioneers, in order to clear what the advanced body had opened for the guns and stores that were to move under cover of the rear division. While we were thus engaged, an unremitting rain, extremely unusual at that season, commenced. The ravines were filled with water, the paths became slippery, the bullocks lost their footing, and the troops were obliged to drag the guns and carriages across the whole forest. . . . The rains continuing fourteen days without intermission, the passage through the forest became daily more distressful, and the troops were exposed in their whole progress, without the possibility of pitching tents, or of procuring for them either cover or convenience. Colingoody is fifteen miles from Palghautcherry, and the road lies entirely through rice grounds, with intersecting ridges covered with cocoa and other trees. The water and embankments necessary for the cultivation of rice render it difficult for guns to pass, and impracticable for cavalry to act. As soon as a sufficient force got through the wood (on the 2nd November), the advanced corps moved to the bank of the Paniani river, within random shot of the works of Palghautcherry, where we took a secure position, and prepared to invest the place. . . . On the 4th November, the main body of the troops, not including the rear division, arrived at our position on the river, which we passed the next day, and encamped about two miles east from the fort, across the great road that leads from Coimbatore. The engineers' stores arrived, and a post for them was established, where all the preparations for a siege were collected. Our next object was to circumscribe the besieged and accelerate our approaches; with this view we occupied the pettah, or open town, on the east and north faces of the fort; and on each of these faces carried forward an attack."

During the whole period of the approaches, and in the construction of the trenches, parallels, and batteries, the besieged kept up a continued fire on the covering and working parties. "The battering train and stores, however, under cover of the 4th Brigade, reached the encampment on the 9th November, after a succession of toil that would appear incredible if recited in detail.

The fort was found to be covered by a glacis, with a good covert-way ; it had a very broad and deep ditch, completely reveted, a large berme, and a very strong commanding rampart. The figure of the fort was nearly quadrangular ; the dimensions of its faces 528 feet by 432 ; each angle was defended by a capacious round bastion with nine enbrasures, and a bastion of a similar construction on the centre of each curtain. It had only one entrance, passing through three gateways, mounted a great number of guns upon the works, and contained a garrison of 4000 men. On the 13th November, fire was opened with twelve guns and four howitzers, from two batteries at 400 yards distance from the east and north faces of the fort, and before sunset the defences were so much damaged that the fire of the besieged considerably abated. On the night of the same day, Captain Maitland, with a part of the four flank battalions, took advantage of a heavy rain to drive the besieged from the covert-way. Being so fortunate as to succeed, he pursued them within the first and as far as the second gateway, where he was stopped, but maintained his ground with great spirit and ability until a reinforcement arrived. This mode of attack so much alarmed the enemy that they called out for quarter, and put us in possession of a fort capable of making a long and desperate resistance. We found 50,000 pagodas (or about 1,75,000 rupees) in the place, besides a very large supply of grain, guns, powder, shot, and military stores. . . . Our loss on the occasion was extremely inconsiderable." The treasure was divided amongst the troops in consideration of their necessities. The Killedar (commander of the fort) and the garrison, with their side arms and private baggage, were set at liberty, and every effort was made to put the place in the best state of defence, and to prepare every department of the army for more important operations.

On the surrender of Palghautcherry, the place was garrisoned by the 19th Battalion, with a few Europeans and some irregulars.

The army then marched to Coimbatore, which was reached on the 26th November, "having been annoyed with rockets, on the march, by a very large body of Mysore Horse." Finding there was no glacis, a battery was erected; but before a breach had been effected, the Killedar surrendered. A great quantity of ammunition, stores, and grain were found in the fort; and the adjacent grounds were covered with abundant crops.

Fullarton then made preparations for an advance against Seringapatam, and sent Captain Maitland with the flank brigade to Dindigul and Tanjore, in order to bring supplies, etc., from the southern garrisons. The third brigade was detached via Palghautcherry to Cochin, where a considerable stock of arrack, stores, and ammunition had been prepared. The main body of the army remained at Coimbatore, ready to oppose the enemy. No further offensive operations were, however, undertaken against Tippoo, for when Fullarton was about to advance towards Seringapatam, with the object of either capturing that place in the absence of Tippoo, who was then besieging Mangalore, or at any rate creating a diversion in favour of the Mangalore garrison, he received orders, on the 28th November, to restore all the places he had captured, and to retire within the limits possessed by the British on the 20th July preceding. This was preliminary to negotiations for peace, which were proceeding between the British and Tippoo. It therefore became necessary for the army to return to Palghautcherry, where the troops received grain to last them as far as Trichinopoly and Madura, a distance of 200 miles. On the 28th December the army advanced towards the southern countries; and at the same time three battalions, under Captain Wheeler, were detached to escort the stores from Cochin, with directions to evacuate Palghautcherry, and to join the army by the route of Annamully and Pulney, close to the mountains. On the second march the army was visited by the missionary, Mr. Swartz, to whom was furnished a large escort for the purpose of proceeding to Seringapatam to negotiate with Tippoo.

1784

On the 4th January 1784 the army arrived at Ayryacotta. Colonel Stuart, with the main division, proceeded to Caroor, on the borders of the Trichinopoly country; Colonel Forbes, with a strong force, remained in the districts of Darapooram and Dindigul; while Colonel Kelly, with another division, advanced to Covanore, on the borders of Madura, and the Marawas. This distribution was made with the object of relieving the country from the burden of supporting too large a proportion of troops. At the same time the divisions were conveniently stationed for concentration in the event of a renewal of the war against Tippoo Sultan. On the 11th March 1784 the treaty was signed, and all the places captured by Fullarton were given up, with the exception of Dindigul, which was held pending the release of the English prisoners.

On the 1st April, Colonel Forbes's Division moved to Dindigul, and Colonel Stuart fell back from Caroor upon the province of Trichinopoly. Whilst stationed at Dindigul, "the troops in that quarter suffered a continuation of fatigue, and were obliged to march seventy miles to the head of the Outumpollum valley, to receive the grain necessary for their current subsistence." By the end of May it became impracticable to support so large a force in the Dindigul country. A strong garrison was therefore left in the fortress; the division was withdrawn towards Madura, and three battalions detached to the Tinnevelly country.

The following is an interesting description by Colonel

Fullarton of the financial state of the officers during the late war:—

"Many officers were obliged to sell their furniture and wearing apparel, in order to procure a scanty subsistence; while others could not possibly find means of appearing as became their station. If a pittance of their arrears was to be advanced, it often came attended with circumstances so singularly disreputable, that nothing short of penury could justify the offer or acceptance; if in Company's bonds, they were hardly negotiable; if in Bengal bills, the holders of them lost thirty, forty, or fifty per cent.; and if the payment took place in an out-garrison, the discretion or caprice of the paymaster alone determined the mode of payment. Needy officers at the mercy of such a superior have frequently submitted to receive a month's arrears in rice, teas, wines, and other merchandise."

Whilst with the 1st Division in Dindigul, Colonel Fullarton made the following report regarding the good conduct of the native troops under his command:—

"The troops have carried their provisions on their backs from Palghautcherry to this place, and have enough remaining to subsist them as far as Madura, being nearly two hundred miles. I mention this circumstance, my Lord and Gentlemen, as a proof of the willing spirit of your sepoys in this quarter, who have borne all their hardships with alacrity seldom equalled, and never surpassed."

On the break up of Fullarton's army in the spring of 1784, it is not certain at which of the three stations, Caroor, Dindigul, or Coranoor, the 13th Battalion was quartered, but shortly afterwards the regiment was moved to Trichinopoly. In October 1784, the distinction between "Carnatic" and "Circar" Battalions being abolished, the 13th Carnatic Battalion became the 13th Madras Battalion.

1785 During 1784 and 1785 Government were in the greatest straits to find funds wherewith to pay the army, which was many months in arrears; and the privations of the sepoys, in consequence, were most lamentable. Notwithstanding the extreme severity of the service, and the fact that they were serving alien masters, they resisted the numerous offers of the Mysore princes, and often escorted remittances for stores, although unpaid and famished for want of food — examples of military fidelity never excelled in the annals of any nation. Promises of payment were not fulfilled, and discontent was reaching the most alarming proportions, when an ill-timed order, directing the reduction of fourteen battalions, precipitated matters, and the 32nd (one of the battalions to be disbanded), thinking that the payment of their arrears might be indefinitely postponed, mutinied at Trichinopoly. The 33rd followed suit, but both battalions submitted quietly a few days afterwards. The Bombay Government came to the assistance of Madras, and signified their intention of forwarding nine lacs of rupees; but before this remittance became available, the 13th Battalion, unable any longer to procure the necessaries of life, mutinied at Trichinopoly on the 14th January 1786. This battalion was described by Brigadier Horne (who commanded at Trichinopoly) as a corps "hitherto distinguished for regularity and obedience, and who, it is with pungent grief I acknowledge, have been driven by a series of want and unparalleled hardships to a state of desperation." The 13th Battalion took possession of the rock, and of the wall which surrounded it, thus ensuring the command of the place. Brigadier Horne, on hearing of this, assembled the rest of the garrison, consisting of 50 men of the Madras European Regiment, 30 artillerymen, a few cavalry and native gunners, and 90 of the 20th Battalion, with which he took up his position at the main guard, to watch the mutineers until the 78th Regiment could come up from Tanjore. In the meantime, a sum equal to pay for one month was offered to the 13th, with the promise of two months more at a short date, but this offer was rejected with indignation. In fact, as had been previously reported, the men of the native corps had lost confidence in the promises of Government, and the 13th assigned the following reasons for their conduct:—

1786

"They said they had done their duty with credit and reputation during the course of the war; that they had submitted to its hardships, in common with the rest of the army, with cheerfulness and alacrity; that they had not murmured, however severe upon them, at the method in which their old arrears were adjusted, because it was general; but that they had been long mortified at the distinction shown to the troops in the neighbouring garrisons by more regular payments, and in a specie not liable to fluctuation; that, however, they imputed this at first to accident, or the distresses of Government. Under that idea they fought with difficulties, sold their little superfluities, and, as wants increased, parted with their necessaries; but that, as insupportable misery came upon them, stifling their cries of nature, and the tenderest feelings of humanity, they were reduced some of them to the dire necessity of consigning their offspring to slavery to preserve their existence. That, in the midst of these sufferings, they saw the troops at Tanjore cleared off for December, particularly the 10th Battalion recently from this station, which had, on its arrival there, received payment up to January, and the 20th Battalion, lately from Madura, for November, while they were offered pay for August, one-half of which was commuted into paddy at an advanced price, with an assurance only of two months more, and without a prospect of being brought on a level with their brother soldiers in their vicinity; that, unconscious of such unmerited distinction, unable to struggle longer with misfortune and contempt, they were resolved to do themselves justice, and they ultimately insisted, as the only terms of accommodation, on a total clearance of all current demands."

Colonel Wilson states that all the garrison guards, as well as a portion of the main guard, were furnished by this battalion, and although they signified their concurrence with their comrades on the rock, in standing out for payment of the arrears which had accrued since the termination of the war, they remained steadily at their posts, performing their duty as usual. Matters remained in this state until the afternoon of the 15th, when the principal officer of the Nawaub came forward with money sufficient to defray two months' pay for the native corps, and the battalion having consented to receive this, on the condition of the speedy discharge of the remaining current arrears, left the rock and returned to barracks about midnight.

The chief reason for the pay of the army having been allowed to get so much in arrears, appears to have been a misconception on the part of the Government of the risks they thereby ran, and a mistaken idea that the development of the country had priority of claim, but the Court of Directors issued the following instructions in 1785:—

"We cannot conclude this despatch without directing your serious attention to the large sum of arrears due to the army. Every possible exertion must be made to diminish them without delay, and all other considerations must yield to that object. The pay of the soldier ought never to be in arrears, and while there is a single pagoda in your treasury, he must be paid in preference to every other article of expenditure."

For the next three years the 13th Battalion, consisting of eight companies—the strength having been reduced from ten to eight companies on the 20th May 1786—appears to have remained at Trichinopoly.

In May 1786 the native troops were divided into five brigades, the 13th Battalion, with the 1st, 6th, 10th, 20th, and 23rd, formed the 1st Brigade, under Lieutenant-Colonel Thomas Bruce, with headquarters at Tanjore, the rest of the brigade being stationed at Trichinopoly and Warrioor. On the 10th March 1787, the 7th took the place of the 10th in the 1st Brigade, which was ordered to garrison Trichinopoly, Tanjore, Madura, Palamcottah, and other places south of the Coleroon.

1787

In April 1789, Government having decided to compel the Rajah of Travancore to contribute towards the up-keep of a subsidiary force, the 13th Battalion, in company with the 10th Battalion and a detachment of artillery—the whole under the command of Captain Knox of the 10th Battalion—marched for Travancore, and were stationed in the island of Vypeen. During this year a dispute arose between Tippoo and the Rajah of Travancore as to the ownership of Jaycottah and Cranganore, which the latter had just purchased from the Dutch. Tippoo demanded their surrender, and the Travancore Rajah refusing to give them up, Tippoo, on the 29th December 1789, attacked the Travancore lines, but was repulsed with great loss. Having received reinforcements, however, from Mysore, he renewed his attack, and carried the lines by storm on the 15th April 1790. Tippoo had made himself master of nearly the

1789

1790

whole of Travancore, with the exception of the island of Vypeen; when, hearing that the English were preparing to send an army to the assistance of the Rajah, he caused the ramparts of the lines to be demolished, and retired with his army into Mysore. Meanwhile the 10th and 13th Battalions had taken no part in the operations. They were joined in the island of Vypeen, on the 24th April 1790, by H.M.'s 75th Regiment, and a detachment of the Bombay Army, under Lieutenant-Colonel Hartley, who assumed command of the whole.

The Madras Army, under the command of Major-General Meadows, having assembled at Trichinopoly on the 24th May 1790, marched for the west two days afterwards. Coimbatore was taken possession of on the 21st July without opposition, and on the 22nd August, Dindigul was captured after severe fighting. Palghautcherry surrendered to Colonel Stuart's detachment on the 22nd September. The 13th Battalion had taken no part in all these operations, but the regiment must have arrived at Palghautcherry shortly afterwards with the force under Colonel Hartley, who, leaving the Bombay troops to garrison that fort, sent the 10th, 13th, and 14th Battalions—all under Captain Knox—to Coimbatore, where they arrived just in time to prevent that place again falling into Tippoo's hands, who, with 4000 men and a large train of artillery, had descended from Mysore about the middle of September, and was manœuvring to catch the British forces, divided as they were, at a disadvantage, and crush them in detail.

About this time the several divisions under Floyd and Stuart rejoined the main army under General Meadows, but nothing decisive occurred until the beginning of November, when Tippoo, hearing that the Bengal contingent under Maxwell, designated the

"Centre Army," was advancing, crossed the Cavery, leaving a cavalry screen to deceive Meadows, and marched against Maxwell; but when he fell in with the Centre Army at Caverypatam, his heart failed him, and he did not venture to attack. Meadows did not discover Tippoo's movements until the 8th November, when he crossed the Cavery near Erode in pursuit, and on the 17th November 1790 effected a junction with the Centre Army at Poolanhully, about twelve miles south of Caverypatam.

On the 20th November 1790 the army was re-brigaded, and the 13th Battalion, with the 4th Madras European Battalion and the 27th Madras Battalion, formed the Infantry Brigade of the advance under Major Gowdie.

Tippoo now marched for Trichinopoly, and, evading all attempts of the British to bring on a general engagement, eventually encamped near Pondicherry. General Meadows marched to Vellout near Madras, where Lord Cornwallis assumed command of the army on the 29th January 1791. On the 5th February the army marched towards Vellore, and on the 11th the whole army was concentrated at that place. On hearing of this movement, Tippoo immediately hastened back to Mysore, but arrived too late to oppose the British, for Cornwallis entered Mysore on the 17th February by the Mugly Pass. He captured Colar on the 28th, Ooscottah on the 2nd March, and on the 5th of the same month arrived before Bangalore. On the 6th, the Cavalry Brigade, having sighted the rear of Tippoo's army when making a reconnaissance, charged, and, pursuing too far, got into difficulties, and had to beat a precipitate retreat. Seeing the state of affairs, Gowdie's Brigade moved up from the swamp where it had been posted, and advanced with its guns to a position on an eminence which commanded the only line for retreat

1791

or pursuit. This enabled the cavalry to rally in their rear, and, opening fire, they soon cleared the field. The latter part of these operations was distinctly visible from the encampment, and Lord Cornwallis immediately proceeded with a division of the army in support. Having arrived at the swamp from which Major Gowdie had advanced without orders, he met, considerably after dark, the cavalry now formed, followed by the infantry and guns, and the whole returning in perfect order to the camp. Major Gowdie, who was guilty of a distinct disobedience of orders, received and deserved nothing but praise. The 13th Battalion in this affair lost three killed and one wounded, the remainder of the brigade losing one man killed. Although the affair terminated favourably for Tippoo, he did not think it advisable to remain on the ground now known to the enemy, but during the night moved six miles further west to Kingeri, leaving 8000 men for the defence of the fort, and 2000 regular infantry and 5000 peons for the defence of the pettah.

The information obtained during this reconnaissance determined Lord Cornwallis to commence the siege of the Bangalore Fort from the north-east, where he was already encamped. This fort had been entirely rebuilt with strong masonry, and the pettah was surrounded by an excellent ditch, planted with impenetrable and well-grown cactus. In order to obtain supplies, and to cover the operations of the siege, dispositions were made on the 7th February to capture the pettah. One European and one native regiment, supported by an equal reserve, traversed the ditch under cover of the thorny defence, and carried the flêche at the point of the bayonet. Whilst the guns were advancing to force the gate, which had been strengthened inside with masonry, the enemy, from the turrets of the gateway, poured a destructive fire of musketry and rockets on the columns. Two ladders would probably have saved many lives, but there was not one in camp; and, after a long delay in making a practicable opening in the gate, which the troops bore with the greatest steadiness and patience, the pettah was at length carried. Tippoo moved from Kingeri with his whole force for its recovery, and despatched the main strength of his infantry by a route concealed from view to the pettah, with positive orders to recover it at all risks. Lord Cornwallis, however, strongly reinforced the pettah, and the Europeans, discontinuing their fire, fell to with the never-failing bayonet. In a contest for the possession of the streets, this mode could neither be evaded nor withstood, and, after severe fighting, the enemy were driven out of the pettah with a loss of about 2000 killed and wounded, the British losing 26 killed and 103 wounded.

"Few sieges have ever been conducted under parallel circumstances; a place not only not invested, but regularly relieved by fresh troops; a besieging army, not only not undisturbed by field operations, but incessantly threatened by the whole of the enemy's force. No day or night elapsed without some new project for frustrating the operations of the siege; and during its continuance, the whole of the besieging army was acocutred every night from sunset to sunrise."

By the 20th, Tippoo perceived distinct indications of an early assault on the fort; and on the morning of the 21st drew up his army on the heights to the south-west, to protect an advanced body with heavy guns, who were endeavouring to explode and destroy the trenches. Lord Cornwallis struck his camp, and drew up his forces as if to attack, though he had arranged to make the real assault during the night. "It was a bright moonlight night, eleven was the hour appointed, and a whisper along the ranks was the signal for advancing in profound silence; the troops were close up to the fort before the garrison took the alarm, and just as the serious struggle commenced on the breach, a few men climbed up the rampart on the left flank of the defenders, where they coolly halted to accumulate their numbers, till sufficient to charge with the bayonet." Once established on the ramparts, the flank companies proceeded as told off, by alternate companies to the right and left, the enemy disputing every inch of ground, until they met over the Mysore Gate. Separate columns then descended into the body of the place, and within an hour all opposition had ceased. On ascending the breach, a heavy column was observed advancing to attack the assailants in flank and rear, but they were repulsed with great slaughter by the troops reserved for this contingency, which had been foreseen. A similar column on the right had been dispersed, at the commencement of the assault, by a body appointed to draw off the enemy's attention from the breach. At the moment the flank companies met over the Mysore Gate, another column was perceived advancing along the sortie to enter and reinforce the garrison, but a few shots from the guns on the ramparts announced that the place had changed hands. The carnage had been severe, but unavoidable, particularly in the pressure of the fugitives at the Mysore Gate, which at length was completely choked; upwards of 1000 bodies were buried, but the number of the wounded was not ascertained. The British lost 103 killed and wounded. Between the 8th and 21st March 1791, the Madras Battalions employed lost 32 killed and 35 wounded.

After this important capture, forage was urgently required for the preservation of the surviving cattle, but Lord Cornwallis could not quit the vicinity until such temporary repairs had been made at the breaches as should place the fortress beyond the immediate risk of a *coup de main*. He moved, however, at daylight on the 22nd February, from the exhausted and horribly offensive encampment which had been occupied during the siege, to the west of the fort. After effecting the repairs above stated, and making preparations for the siege of Seringapatam, he moved, on the 28th February, in a northern direction towards Deonhully; despatching, on the preceding night, a battalion to prevent, if possible, the destruction of forage in the villages adjacent to the intended encampment. Tippoo had moved on the same morning in the direction of Great Balipoor. The roads on which the hostile armies were marching crossed each other diagonally. The Battalion perceived at daybreak the enemy's columns of march crossing his front, and had no alternative but to take post. Tippoo, conceiving that he saw the British advance guard, quickened his pace to clear it. In the meanwhile, the real advance guard, on ascending an eminence, saw the greater part of the enemy crossing diagonally at a distance of three miles, and also took post. On ascertaining these circumstances, Lord Cornwallis advanced with all possible expedition. The cattle, reduced to skeletons, were scarcely able to move their own weight; the soldiers, European and native, everywhere spontaneously seized the drag ropes and advanced the guns frequently at a run. The British artillery successively dispersed them at every stand they attempted; the infantry continued the pursuit until they were compelled to break into several columns on different roads to effect their retreat. The British Army halted after a march of twenty miles, being double the distance it had been deemed possible to drag the cattle along, and pitched their camp in a situation surrounded with excellent dry forage. Tippoo meanwhile collected his scattered columns near Great Balipoor on the same night, after a march of twenty-six miles; but not considering his position, about eleven miles from the British encampment, to be sufficiently distant, he resumed his march after a few hours' rest towards Shevagunga.

The object of Lord Cornwallis's movement was to effect a junction with the corps of cavalry prepared by Nizam Ali to serve with his army. During the march, the forts of Deonhully and Little Balipoor surrendered without opposition. Tippoo's activity against the British Army was skilfully displayed in the dissemination of false intelligence. After a march of about seventy miles north, Lord Cornwallis remained stationary for five days; and, being deceived by reports which induced him to abandon the hope of forming the junction, he started to march south; but on the evening of his first retrograde march he received more correct information, which caused him to resume the northern route, and the junction was formed on the second day afterwards, at Cottapilly. The united bodies then moved south for the purpose of joining a convoy advancing by the passes near Amboor, which was escorted by 4000 men. Tippoo decided to strike at this convoy, but was foiled in his preparatory movements by the superior strategy of his opponent.

"The supplies of provisions and stores which had been collected for the army at Amboor, with a reinforcement of four battalions of sepoys, having joined the camp at Vencatigherry on the 21st April 1791, the Grand Army moved on the 22nd towards Bangalore. The march was productive of nothing worthy of notice, if we except the conduct of our Nizamite allies, who, in some slight skirmishes with the enemy's horse, gave us but little reason to hope for much material benefit from their assistance. Tippoo's army kept at a short distance from us, and his detached parties hovered round us to watch our motions." It was now decided to hasten the conclusion of the war by attempting the siege of Seringapatam, in spite of the great want of transport. Whilst the army was encamped at Bangalore, and the necessary arrangements made for approaching the capital, Tippoo remained in the neighbourhood of Savandroog, from whence he moved to Seringapatam by the shortest road, as soon as he was certain that the British Army was moving to the same point.

"At Bangalore we found that our means of conveyance for provisions and stores were extremely inadequate to the quantity of both which it was thought necessary to convey." Lord Cornwallis appealed to all officers to apply whatever means of carriage they had in their power to command, for the carriage of shot to the place of destination. "Followers of various kinds, chiefly the relations of sepoys, were also prevailed upon to undertake carriage proportioned to their means; even women and boys carrying each an 18-lb. shot; and by these extraordinary expedients the English General was enabled to advance towards his object." In order to transport provisions, as much grain was given gratis, from the stores in Bangalore, to each sepoy as he chose to carry.

Before leaving Bangalore, Lord Cornwallis formed the Grand Army, on the 2nd May, into two wings, the right wing consisting of one brigade of Europeans, with three brigades of Bengal Sepoys on the right, and four brigades of Bengal Sepoys on the left, and the left wing of two brigades of Europeans, with six brigades of Coast Sepoys on the right, and five brigades of Coast Sepoys on the left. The artillery was formed up between the two wings. At about 1000 yards in rear of the wings, the reserve, under the command of Lieut.-Colonel Floyd, was formed up. The reserve was composed of the 7th Brigade, under the command of Major Gowdie (consisting of the 4th Europeans, with the 13th Battalion, under Captain Oliver, on the right, and the 27th Battalion on the left), with the cavalry on the right and left flanks.

Leaving a garrison consisting of 2000 native troops

and 200 Europeans at Bangalore, the army, which up to that time was the largest regular force ever assembled in India, marched on the 3rd May for Seringapatam.

Tippoo, joined by the division from Gooty, took up a strong position in the main road to Seringapatam, named the Cenapatam road, supported by the hill forts of Ramgherry and Sevengherry, where he declared his intention of making a serious stand. The British General had correct intelligence of the advantages of this position, and of the wholesale destruction of forage and grain on that route; and hoped to avoid some of these inconveniences by adopting the more circuitous route of Caunkanhully nearer the Cavery. It was only on his first march, however, that he benefited by this determination; "for from that period forward, not only was every march preceded by a wide conflagration, but every human being on the route was so completely removed beyond the reach of the British Army, that they appeared to be traversing a country of which the population had been utterly destroyed by some recent convulsion of nature." All the inhabitants, with their cattle and movables, had been collected on the island of Sheven Summooder. The exhaustion of the transport cattle daily increased, and large quantities of stores were destroyed because they could not be carried, although a large and increasing proportion was dragged by the troops, in spite of the active pressure of the enemy in rear. To add to the difficulties, heavy rains came on when the army reached Sultanpettah. "The Nizam's troops were so much in dread of Tippoo's cavalry that they never ventured beyond our picquets; and instead of procuring forage and provisions, served only to consume the gleanings of the country, which, had we not been encumbered with their assistance, would have maintained our cattle and followers. The wants of the latter description of people brought on us another misfortune. The sepoys, for whose families no provision had been made, trusting to those supplies they had formerly been accustomed to find by their own industry, had shared with them what had been given for their own subsistence, and instead of having, as was supposed, provisions for ten days remaining in their possession, many were in absolute want. Punishing the men for thus misapplying their stock would not remedy the evil; provisions were therefore served out to them from the public store, at half allowance." Major Dirom, in his *Narrative of the Campaign*, says that each field officer had 40 servants, each captain 20, and each subaltern 10 servants during the campaign, and makes the following remarks on the followers of the native regiments :—

"Early in the war, many of the sepoys were prevailed upon to send back their families, and other arrangements were made for reducing the number of followers; but these measures tended to create desertion and increase distress. In short, no man will carry his family to camp who does not find his convenience and advantage in doing so; no person will pay for servants he does not want, nor will followers attend on an army without pay, who do not earn a living, which they can do by contributing to its support. . . . Sepoys, most of whom are married, have many of them, as well as the followers, their families in camp."

The army reached Arikera, nine miles east of Seringapatam, on the 13th May, when Lord Cornwallis perceived Tippoo's army occupying strong ground about six miles in his front, with their right on the Cavery, and their left along a rugged and apparently inaccessible mountain. Lord Cornwallis resolved to attempt, by a night march, to turn the enemy's left flank; and, by gaining his rear before daylight, to cut off the retreat of the main body of his army to the fort and island of Seringapatam. Accordingly, leaving his camp and stores under a strong guard, he marched some hours before daylight. Retarded, however, by bad roads and bad weather, he had not proceeded above three or four miles when day broke, so he abandoned his original idea, and, attacking the enemy where they stood, drove them back to a position under the guns of Seringapatam. The 13th Battalion did not take part in this latter engagement, as it formed part of the camp and baggage guard with the rest of Gowdie's Brigade. The British Army lay on their arms, nearly on the ground on which the action terminated, and, after the arrival of the tents in the course of the night (the 15th May), encamped just beyond the range of the cannon on the island.

"Our battering train joined us on the 16th May, and on the 18th the army moved to the north-west of the island, and encamped at the foot of a remarkable rock called Yirdimally, on the Milgotah road. The consequences of our want of forage were now severely felt. Our bullocks, in a march of six miles, were quite exhausted, and the guns and stores required all the exertions of the troops to bring them forward. After a day's halt we again moved to Caniambaddy, where there is a ford across the Cavery, about eight miles above the town. This march (on the 20th May) was extremely fatiguing. The heavy guns were almost entirely dragged by the troops, and four battalions were employed on the same duty with the store department. The rains had set in decidedly, and provisions for the followers, and even for the fighting men, were become scarce. The 21st was employed in repairing the ford for the passage of guns; and on the morning of that day, a large detachment of the enemy passed in our view along the south side of the river, on their march towards the place where the Bombay Army was encamped. The season being so far advanced, and our supplies of provisions so deficient, the measure of beginning a siege of such importance as that of Seringapatam would have been absurd."

Accordingly, Lord Cornwallis determined to relinquish the plan of campaign; and on the 22nd May he destroyed the whole of the battering train and heavy equipment. All that had occurred of mortality among the cattle during the siege of Bangalore fell far short of what happened at Caniambaddy, rendering the air pestilential and the surroundings disgusting. "The ration of rice to the fighting men had now for some time been necessarily reduced one-half; the appearance of the sepoys, of whom a large proportion live exclusively on vegetable food, indicated a gradual but very perceptible wasting and prostration of strength. A bleak wind and continued drizzling rain had more than its usual influence on constitutions shaken by other causes." On the 24th May, two brigades crossed the river with a view to induce

Tippoo to recall the detachment he had sent against the Bombay Army. The river rising fast from the unremitting rains that fell for some days, the detachment recrossed, as their situation appeared extremely hazardous.

On the 26th May the army moved from the encampment at Caniambaddy by a route to the northward towards Bangalore, "our slow march bearing strong testimony to the reduced and weak state of our means and conveyance; even the light field-pieces required the assistance of the battalions to which they are attached to move them forward."

"At the close of this march, a body of horse appeared on our baggage flank; and while some corps were forming to oppose them, some men, to our great surprise, rode into our line with information that this party belonged to the van of the Mahratta army, which was distant eight miles, on its way to join us. This was the first intelligence we had of their approach; and although forty letters had been despatched from them to apprise the Commander-in-Chief of their intentions, yet so closely was our camp surrounded by Tippoo's horse, that not one of the number had reached us. Two days after the receipt of this intelligence we moved towards Milgotah, on the road to which the united Mahratta armies encamped close to ours. They brought with them a large bazaar, from which the British Army obtained supplies. It having been determined to move to a more plentiful country, till we should be prepared and the season favourable for renewing the attack upon the capital, we decamped on the 6th June from the neighbourhood of Seringapatam, and proceeded by short marches to the north-east, making frequent halts." The army passed the vicinity of Hooliordroog, a small impregnable rock with a town at its foot; the town was easily carried, and the garrison capitulated. Shortly afterwards this fort was razed to the ground. Other small hill forts were taken possession of on the way. Savandroog was reconnoitred during the march, near which place the army halted for some days. In consequence of its great strength, no attempt was made to capture it. The army arrived at Bangalore on the 11th July 1791. Foraging parties were sent out and supplies brought in by the troops, in spite of many scattered bodies of the enemy who greatly harassed them. At Bangalore "the gentlemen of our line had an opportunity of supplying themselves with liquors, tea, etc., luxuries which in the Grand Army had for a considerable time been almost unknown."

The Policade Pass afforded the easiest communication with the Carnatic; and one of the most convenient issues for the sudden incursions of the enemy. It was commanded by several forts, of which Oossoor and Rayacottah were the chief. "On the 15th July 1791 the army moved from the neighbourhood of Bangalore, and marched towards Oossoor, through a fertile country, beautified with chains of tanks for the culture of the low grounds, and with numerous villages and small forts; which, surrounded with trees, crowned and adorned every eminence. The 7th Brigade, commanded by Major Gowdie (of which the 13th Battalion of Madras Infantry still formed part), having been sent a march in front of the army, reached Oossoor on the 15th July." Other columns had also started to reduce other forts west of Bangalore, and establish posts throughout the country. On Gowdie's approach, the enemy were preparing to abandon the place, but "being dilatory on all occasions, the unexpected approach of this brigade forced them to beat a precipitate retreat. They spiked the guns, burnt the carriages, and, as they went off, fired a mine, which blew up one of the bastions; but a train they had laid to the magazine luckily did not take effect, so that the damage done to the place was not great; and both a quantity of powder and other ammunition, and a considerable store of grain which the garrison had not time to remove, was found in the fort." Tippoo had lately made great exertions to improve the defences of this important place, but fortunately they were not so far advanced as to render it tenable in the opinion of its defenders.

The hill forts of Auchittydroog, Neilgherry, and Ruttungherry surrendered a few days afterwards without resistance.

From Oossoor, where a detachment of Bengal troops had been left in garrison, Major Gowdie's Brigade proceeded against Rayacottah on the 19th July. This place consisted of two forts, one at the bottom, the other at the top of a precipitous rock, and was garrisoned by eight hundred men. The Killedar refusing to surrender, the Major at daylight the next morning blew up the gate and attacked the lower fort, which enclosed the pettah, hoping to carry the rock by entering with the fugitives. The stormers were led by Captain Oliver of the 13th Madras Battalion. "He soon carried the lower fort by assault; and, pursuing the fugitives, got possession of the first two walls, which form a sort of middle fort between the lower fort and that which defends the summit of this stupendous rock. The place was known to be too strong by nature to be reduced, if the garrison were resolute in its defence; Major Gowdie was therefore directed to return to the army, if it was not given up on the first attack; but, having made a lodgment on the hill, where the troops were covered from the fire of the upper fort, and having reason to think the garrison were intimidated, he requested leave to continue the attack." On the 22nd July he was reinforced by one battalion and two guns. The daring conduct of the assailants, and the appearance of the army, soon produced the desired effect upon the mind of the Killedar, who proposed a parley; and, on condition of security to private property, and leave to reside with his family in the Carnatic, he surrendered this "lofty and spacious fort, so strong and complete in all respects, that it ought to have yielded only to famine and a tedious blockade," on the 22nd July.

"On the 29th July the army returned above the pass from Rayacottah, and continued in the neighbourhood of Oossoor to cover the convoy from Amboor, now ordered to come up by this route." Kinchillydroog, Oodiadroog, and other small hill forts by which the Policade Pass was defended, submitted shortly afterwards, and were either garrisoned or destroyed.

The approaches from the Carnatic having been thus secured, Major Gowdie's Brigade was sent against Nundydroog and other forts to the north of Bangalore, which interrupted the communication between that place and the Nizam's army near Gurrumcondah. On the 14th September, Major Gowdie was joined within nine miles of Rahmanghur by Captain Read's detachment and a party of Bengal Artillery. He immediately proceeded to reconnoitre, and summoned the fort at that place to surrender. As the Killedar refused to come to terms, a plan was immediately fixed on for the siege. "On the 17th, batteries constructed within a few hundred yards of the walls were ready to open fire, when the Killedar, alarmed at the rapidity and vigour of the attack, proposed to give up the place on terms which, being refused to him, the batteries opened on the afternoon of the 17th, when a few rounds of shot and shells determined him to surrender at discretion; and in the evening, the Major, without having sustained any loss, was put in possession of a place of great strength and importance."

The only fortress which made any considerable

resistance was Nundydroog, before which the Brigade arrived on the 22nd September. The pettah was attacked and carried the same day, but the fort itself was found to require larger reinforcements and more extensive measures. After forcing the pettah, Major Gowdie, having examined the northern face, and finding it unassailable in that quarter, made a circuit to the west, and finally invested the place on the 27th. The fort was situated on the summit of a mountain about 1700 feet in height, of which three-fourths of the circumference was absolutely inaccessible, and the only part practicable for assault was guarded by two excellent walls, and by an outwork which covered the gateway and provided a flank fire. A road was cut, and the guns dragged with infinite difficulty to the top of an adjacent hill; but after a battery was erected, the guns were found to be too distant to reach the defences of the fort.

"There was no alternative but to abandon the attack, or attempt to work up the face of this steep and rugged mountain to within breaching distance of the fort. This arduous undertaking was adopted, rather than leave a post of such consequence in the possession of the enemy, and encourage them by an instance of our troops being foiled in the attack of a fortified place, which had not yet happened during the war. The exertions required to form a gun road and erect batteries in the face of this mountain, surpassed whatever had been known in any former siege in India; and such was the steepness of the ascent, that the battering guns could not have been drawn up without the assistance of elephants. During a fortnight that the troops were employed in this last arduous work, a continual fire was kept up on them from the fort. The cannon shot, directed from so great a height, seldom took effect; but they were severely galled by the jingal, a species of wall piece, which threw with precision, to a great distance, a ball of considerable size. The batteries formed, two breaches were made, one on the re-entering angle of the outwork; the other in the curtain of the outer wall. The inner wall, at a distance of eighty yards, could not be reached by the shot. The Governor still refused to surrender. The breaches being reported practicable to the Commander-in-Chief, who had now moved the army to the immediate vicinity, it was determined to attempt to storm the inner wall by escalade, and if this should fail, to make a lodgment behind a cavalier between the walls, and thence attempt to breach it. The assault was given by clear moonlight on the morning of the 19th October. The fort was illuminated with blue lights, and a heavy fire was opened. Masses of granite were rolled down the rock with tremendous effect, and although the garrison was perfectly alert, the ardour and rapidity of the assailants surmounted every obstacle, and they pressed the fugitives so closely as to prevent their effectually barricading the gate of the inner rampart. It was forced after a sharp conflict, and the place was carried, with the loss in the assault of only two killed and twenty-eight wounded (chiefly by the stones rolled down the rock); and in the whole siege, 120 casualties."

"Nundydroog, defended by seventeen pieces of cannon, chiefly iron guns of a large calibre, improved by its late works, and well garrisoned, was thus taken by regular attack in the course of three weeks, although of such strength that it was not yielded to Hyder by the Mahrattas till after a tedious blockade of three years."

The following is an extract from a General After Order published by the Commander-in-Chief:—

"CAMP, *19th Oct.* 1791.

"Lord Cornwallis, having been witness of the extraordinary obstacles, both of nature and art, which were opposed to the detachment of the army that attacked Nundydroog, he cannot too highly applaud the firmness and exertions which were manifested by all ranks in carrying on the operations of the siege ; or the valour and discipline which was displayed by the flank companies of H.M.'s 36th and 71st Regiments ; those of the Madras 4th European Battalion, the 13th Bengal Battalion of Native Infantry, and of the 3rd, 4th, 10th, 13th, and 27th Battalions of Madras Native Infantry, that were employed in the assault of last night, and which, by overcoming all difficulties, effected the reduction of that important fort.

"His Lordship is highly sensible of the zealous and meritorious conduct of Major Gowdie in the command of that detachment, both at the attacks of Raymanghur and in carrying on the arduous operation of the siege of Nundydroog, for which the Major will be pleased to accept his best acknowledgments. The whole of the officers and soldiers who composed that detachment appear likewise to be justly entitled to the strongest expressions of his approbation."

The neighbouring fort of Cummuldroog surrendered immediately after the capture of Nundydroog.

Between Bangalore and Seringapatam lies a chain of hills, thickly covered with wood, extending from the vicinity of Bangalore to the river Madoor. This difficult country, which of itself formed a strong barrier to the capital of Mysore, was studded with forts, of which some, particularly Savandroog, were of extraordinary strength. It offered such advantages to the enemy for interrupting the communications with Bangalore, when the army should advance to Seringapatam, that the Brinjarries, who had contracted to supply large quantities of grain from Bangalore, would not undertake to deliver it beyond Savandroog, if that fortress remained in the enemy's hands. Lord Cornwallis resolved to make an effort to gain possession of this important and formidable post. It is a vast mountain of rock, computed to rise above half a mile in perpendicular height, with a base of eight or ten miles in circumference, surrounded by a close forest several miles in depth, having its natural impenetrability augmented by thickets of planted bamboos. "A narrow path, cut through the jungle in a winding direction, and defended by barriers, served as the only approach to the fort. The natural strength of the mountain had been increased by enormous walls and barriers which defended every accessible point. To these advantages was added the division of the mountain by a great chasm into two parts at the top, on each of which was erected a citadel, the one affording a secure retreat, though the other was taken, and by that means doubling the labour of reduction."

About the middle of December the 13th Battalion, under Captain Oliver, joined Lieutenant-Colonel Stuart's force, consisting of H.M's. 52nd and 72nd Regiments, the 14th and 26th Bengal Battalions, and the 6th Madras Battalion, which was encamped about three miles to the north of that side of the rock from which it was proposed to carry out the attack; while the Commander-in-Chief made that disposition of the rest of the army which seemed best adapted to cover the besiegers, and secure the communication. The labour of cutting a way through the dense jungle, and transporting heavy guns over the rocks and hills which intervened, was excessive in the extreme. "The closeness of the surrounding hills and woods had rendered this fortress as remarkable for its noxious atmosphere as its strength." Its name signified, literally, the rock of death. Tippoo congratulated his army upon the siege; at which one-half, he said, of the British Army would be destroyed by sickness, the other by the sword. The confidence of the garrison in the strength of the place had this good effect, that it made them regard the approach of the besiegers as of little importance, and they were allowed to erect their batteries without any further opposition than the fire from the fort. The first battery opened fire on the 17th December, and three days afterwards the breach was practicable. The jungle was now of advantage, for, growing close up to the very wall, the troops were able to scramble up by the crevices and rugged parts of the rock unseen, and made a lodgment within twenty yards of the breach.

The 21st December was the day chosen for the assault. At an hour before noon, on a signal of two guns from the batteries, the troops advanced to the breach. The enemy, who had descended for the defence of the breach, when they beheld the British advancing, were seized with a panic, and there was little difficulty in carrying the eastern section. The danger was lest the flying enemy should gain the western summit, which, from the steepness of the approach, and the strength of the works, might require a repetition of the siege. Fortunately the enemy impeded one another in the steep and narrow path up which they crowded to the citadel, and some shot, which opportunely fell among them from the batteries, increased their confusion. The British entered the barriers along with the enemy, of whom about 100 were killed on the western hill, and several fell down the precipices endeavouring to escape. The prisoners taken were few. The garrison had consisted of 1500 men, but a great part of them had deserted during the siege. The place was carried in less than an hour, without the loss of a man.

The following Order was published by the Commander-in-Chief:—

> "Lord Cornwallis thinks himself fortunate, almost beyond example, in having acquired, by assault, a fortress of so much strength and reputation, and of such inestimable value to the public interests, as Savandroog, without having to regret the loss of a single soldier on the occasion. He can only attribute the pusillanimity of the enemy yesterday to their astonishment at seeing the good order and determined countenance with which the troops who were employed in the assault, entered the breaches and ascended precipices that have hitherto been considered in the country as inaccessible."

The 13th Battalion was left to garrison the fort at Savandroog, and remained there till March 1792, when, terms of peace having been settled with Tippoo, the

1792

army returned to quarters. By virtue of the treaty, the British and their allies obtained one-half of Tippoo's dominions, three crores and thirty lacs of rupees, and the release of all prisoners.

The prize and other money ranged from £308, 10s. for a captain, to £5, 18s. 1d. for a sepoy.

The services of the native troops during the campaigns against Tippoo were rewarded by a medal bearing on the obverse the figure of a native officer holding a flag, and on the reverse the inscription "for services in Mysore, A.D. 1791–1792," within a laurel wreath surrounded by an inscription in Persian. Six months' batta was also given to all the officers and troops.

The 13th Battalion under Captain Oliver, after quitting Savandroog, went into garrison at Dindigul, where Major Dalrymple with a detachment of H.M.'s 71st was in command.

War having been declared between the British and the French, and Pondicherry having again been captured, an expedition was planned, about the end of April 1794, against Mauritius, held by the French. The native troops intended for this expedition were composed of three volunteer battalions, the 1st Volunteer Battalion being formed of volunteers from the 1st, 2nd, 5th, 9th, and 13th Battalions; but the expedition was abandoned, and the volunteers returned to their respective regiments in July and August of the same year.

1794

On the 29th August 1794, the system of field exercise in force with the Royal Army was introduced into the Madras Army, with the following deviation: "That the Army of the Coast continue to form two deep, the officers and sergeants not posted in the ranks forming a third, to be considered as the supernumerary rank. The corps, however, are occasionally to practise their exercise and movements three deep." Sir Eyre Coote had, in December 1780, issued an order that all corps should parade two deep, an order which, however, was neglected.

1795

During 1795 the Polygar chiefs in the south gave a good deal of trouble, and the 13th Battalion took a prominent part in suppressing them and quieting the country. The following is an extract from Wilson's *History*, describing the good service done by this regiment on one occasion :—

" Early on the morning of the 7th October, Captain James Oliver, with a detachment of the 13th Battalion, surprised and captured Valoidum Naik, the Polygar of Pylney, in his fort at Baulsamoodrum, in the Madura district. This chief having been in rebellion for some time, his capture was considered of so much importance that Captain Oliver and his detachment were thanked in General Orders ; and Boodh Sing, the jemadar of the party, was promoted to subadar, and presented with a gold medal, having on one side the words, 'Courage and Fidelity,' and on the other, 'By Government, 7th October 1795.'"

"The year 1795 saw several discontents in the army of the East India Company; and in one of his letters Sir John Shore says: 'If you were to judge of its temper from the conversation of individuals, you would conclude that the officers were in an actual state of mutiny.' Some new regulations, forming part of a plan originally conceived by Lord Cornwallis, to transfer the Company's army to the King's service, were partly the cause of this. The whole organization of the Indian army was changed. Instead of a single battalion of 1000 men, commanded by a captain, who was selected from the European regiments in the Company's service, with a subaltern to each company, they were formed into corps of two battalions, to which officers were appointed of the same rank and number as in the King's regiment. Matters relating to promotion, pay, and allowances added to the ferment. Among the amendments were regimental rise to the rank of major, increased allowances to the senior officers of the army, addition to the staff of the native cavalry and infantry as regarded their military and medical branches, and of passage-money to subalterns compelled to return home by ill health."

1796 On the 12th and 13th July 1796, the Orders reorganizing the Madras Army were published, by which the existing battalions were formed into regiments of two battalions each. Under this new system the 13th Battalion became the 2nd Battalion of the 3rd Regiment, and the left wing of the 22nd Battalion being formed into three companies, was incorporated into it. The selection of the 13th Battalion as the 2nd Battalion of the 3rd Regiment does not appear to have been made with reference to any connection existing between the battalions, but rather to have been determined by their location at the time, as the 3rd Battalion (now the 1st Battalion, 3rd Regiment) was stationed in 1796 in Madura, and the 13th Battalion in Dindigul. It was not long after this reorganization took place, that detachments from both battalions of the 3rd Regiment were employed on service together; for in the autumn of 1796 a force under the command of Major Haliburton, including 180 men of the 1st Battalion, 3rd Regiment, and 300 men of the 2nd Battalion, 3rd Regiment, proceeded against the Polygars, near Dindigul. The disturbances were soon suppressed, as the detachment met with little opposition, and shortly afterwards the men employed rejoined their stations.

1798 On the 12th July 1798, the officers of the 2nd Battalion, 3rd Regiment, subscribed the following sums to the "Patriotic Contribution Fund," which were remitted to England for the purpose of assisting the British Government in carrying on the war against France :—

Lieutenant-Colonel Oliver	Pagodas	200
Captain S. Cuppage	"	100
" E. M. Gepp	"	100
" P. H. Keay	"	100
Lieutenant Cunningham	"	100
" C. Rand	"	100
" H. Smith	"	50
" H. M'Intosh	"	40
" P. Baynes	"	50
" H. F. Smith	"	20
" S. Fish	"	20
" F. Ahmutty	"	20
Assistant-Surgeon Inverarity	"	50
Total pagodas		950

Towards the end of 1798, intelligence of the invasion of Egypt by the French, having reached India, the Governor-General addressed Tippoo, protesting against his intercourse with the French, and offering to appoint a British Resident to Seringapatam to improve relations between him and the British. Having received an evasive reply, orders for the assembly of the army near Vellore, under General Harris, were given; and to a second communication a similar reply from Tippoo having been received, the army was directed to advance into Mysore. The 2nd Battalion, 3rd Regiment, 1023 strong, was brigaded with the 1st Battalion, 8th Regiment, and the 2nd Battalion, 12th Regiment, under Colonel Roberts, Madras Army, and formed the 5th Brigade, which, with the 1st and 3rd Brigades, composed the right wing under Major-General Bridges, Madras Army.

The following is an extract from the *Asiatic Annual Register* for 1799:—

"*January 29th*, 1799.—The 2nd Battalion, 3rd Regiment, Native Infantry was reviewed by Major-General Floyd. After the review, a very elegant breakfast was given by Colonel Oliver at his garden-house, at which General Floyd and all the officers of the garrison were present. The following is a copy of the General Orders of the 29th December 1798:—

"'Major-General Floyd desires to express to Lieutenant-Colonel Oliver, his officers and men, that he is extremely satisfied with the appearance and performance of the 2nd Battalion, 3rd Regiment, at the review this day; the General considers the corps as very fit to be presented either to friend or foe, and will report to headquarters accordingly.'

"In the afternoon a very splendid entertainment was given by General Floyd to the officers of the 2nd Battalion and several other gentlemen, and the evening passed with the harmony and conviviality which General Floyd knows so well the art of diffusing among his guests.

1799 "On the 1st the battalion marched for Wallajabad; and from the camp at Seringham the officers sent the following letter:—

"'TO MAJOR-GENERAL FLOYD, ETC.

"'SIR,—We, the officers of the 2nd Battalion, 3rd Native Regiment, impressed with a due sense of your polite conduct in a private capacity, and feeling ourselves further gratified by the flattering terms in which the corps was mentioned in G.O., beg leave to return you our sincere thanks. Should it ever be our good fortune to be placed again under your command, we hope to merit a continuance of that approbation.—We have the honour to remain, Sir, with respect and attachment, your faithful humble servants, etc.'

"General Floyd immediately returned the following polite answer:—

"'TRICHINOPOLY, *Jan. 2nd.*

"'TO LIEUTENANT-COLONEL OLIVER AND OFFICERS OF THE 2ND BATT., 3RD REGIMENT.

"'GENTLEMEN,—Your letter of yesterday's date is extremely gratifying; my attentions to you in private were the consequences of your amiable manners; and my public opinion of the corps is founded entirely on the good order, zeal, and alacrity observable through all ranks; and I shall be proud to serve with the 2nd Battalion, 3rd Regiment, on any occasion.—I have the honour to be, Gentlemen, your faithful humble servant,

J. FLOYD.'"

On the 14th February 1799, the army of Madras, some 21,000 strong, marched for the frontier, and was joined on the 20th by 16,000 men from Hyderabad.

The army entered the territory of Mysore on the 5th March, and commenced operations by the reduction of several forts upon the frontier. "Some of these forts surrendered without any resistance, and none of them were defended with spirit." The fort of Oodiadroog surrendered to the 2nd Battalion, 3rd Regiment, under Lieutenant-Colonel Oliver.

The progress of the army was unavoidably slow, owing to the presence of the large amount of ordnance necessary for the siege of Seringapatam; its movements, however, were but little impeded by the enemy, although considerable bodies of horse hovered about its line of march. During the march little or no food or forage could be procured (as Tippoo had everywhere destroyed the villages) to add to the stock conveyed with the army on the backs of the bullocks. According to one account, every tank and pool of water was impregnated with poison of the milk-hedge, large quantities of the branches of which the enemy had treacherously thrown in, so that many horses, bullocks, and in some instances soldiers and camp followers, fell victims to the deleterious infusion.

General Harris advanced, on the 26th March, to a position between Sultanpet and Mallavelly with little opposition from the enemy; but on the 27th March, when the army reached its ground at Mallavelly, Tippoo opened a distant cannonade upon it, which, though at first ignored by the British, ultimately led to a general engagement, in which the enemy was completely defeated, and driven from every position in which they attempted to rally. The British loss on this occasion amounted to sixty-six men killed, wounded, and missing, of which the 2nd Battalion, 3rd Regiment, had three wounded and one missing; while that of the enemy is supposed to have amounted to two thousand killed and wounded.

The following General Order was published in the evening:—

"CAMP MALLEVILLE, *27th March* 1799.

"G.O.—The Commander-in-Chief congratulates the army on the happy result of this day's action, during which he had various opportunities of witnessing their gallantry, coolness, and steady attention to orders.

(Signed) B. CLOSE, Adjt.-Gen."

On the 29th March, General Harris marched towards the river Cavery, where, finding a ford at some distance above the junction of the Cavery and Copany, he immediately crossed the Cavery with part of his army, and occupied strong positions on both banks of that river, at a distance of about fifteen miles from Seringapatam. This movement, so advantageous for the acquisition of supplies, and for facility of junction with the army of Bombay, was made without the least opposition on the part of the enemy.

The whole army having crossed the Cavery on the 30th March, halted near the village of Sovelly on the 31st. On the 1st April, General Harris moved towards Seringapatam, to which place Tippoo had retreated, and on the 5th encamped two miles south-west of that city, having experienced no opposition from the enemy since the 27th March.

On the evening of the 5th, H.M.'s 12th Regiment and the 1st Battalion, 1st Regiment, and the 2nd Battalion, 3rd Regiment, under Lieutenant-Colonel Shawe, were selected for a night attack on an advanced post of the enemy, situated in a ruined village, two thousand yards from the fort and in front of the British left. The attack was successful, and the post occupied by the force. During the advance in the darkness of the night, the 2nd Battalion, 3rd Regiment, fell into confusion. It appears that at this affair Major Colin Campbell of the 1st Regiment was doing duty with the regiment, and was commanding at the time; he met his death whilst attempting to rally the regiment in the dark. Colonel Wellesley with three regiments made a simultaneous attack on a wood near the post attacked by Colonel Shawe, but, being met with a tremendous fire of musketry and rockets on every side, his men gave way, were dispersed, and retreated in disorder. The enemy, retaining the position, "severely galled, with musketry and rockets, the troops posted in the village taken by Lieutenant-Colonel Shawe, during the whole of that night and part of the succeeding day." A renewal of the attack thus became necessary on the morning of the 6th, which was completely successful, putting the British in possession of a strong line of posts, from the bank of the Cavery to Sultanpet, which formed the right of the position. Guns were mounted in all the captured advanced posts.

"On the 2nd May the breaching batteries were opened early in the morning with admirable effect, and before the evening the outer wall was perfectly breached, and the principal rampart considerably damaged." During the night the enemy attempted to repair these, but the fire from the batteries the next day rendered the breach nearly practicable, and the assault was fixed for the 4th May. In order that no extraordinary movement might lead the enemy to expect the assault, which was to be made in the heat of the day, the troops intended to be employed were stationed in the trenches early in the morning of that day. The native troops for the assault were selected from the three Presidencies, ten companies Bengal, eight Madras (to which the 2nd Battalion, 3rd Regiment, contributed its two flank companies, commanded by Lieutenant Arthur Molesworth), and six Bombay. Lieutenant-Colonel Dalrymple commanded the Madras contingent, which was placed in the right attack under Colonel Sherbrooke.

The following are extracts descriptive of the assault:—

"On the morning of the 4th May the batteries kept up an incessant and well-directed fire on the breach and remaining defences of the fort, which was warmly returned by the enemy till noon, when, as usual, their fire slackened. . . . From knowledge of the customs of the natives of India, I judged that during the heat of the day the troops of the garrison would not be apprehensive of an assault, or prepared to make that obstinate resistance which at any other time I might expect to be opposed to our attack. I therefore directed it to take place at one o'clock. The troops passed the rugged bed of the Cavery, which, opposite to the breach, was about two hundred and eighty yards in breadth, exposed to a very heavy fire from the still numerous artillery of the fort, crossed the ditch, and ascended the breach in despite of all opposition from the enemy, many of whom rushed down the slope to meet them. The assailants divided, as they had been instructed to do, at the summit of the breach, and although obstinately resisted by the enemy posted behind a succession of traverses thrown up across the ramparts, particularly on the northern face of the fort, in two hours the whole of the works were occupied by our troops, and the British colours flying in the place."

The troops detailed for the right attack were ordered to move along the southern rampart of the fort; the European pioneers to carry scaling ladders, assisted by forty men of the battalion companies of the leading regiment; the native pioneers to carry a proportion of fascines; the leading companies of each attack to use the bayonet principally, and not to fire except in case of absolute necessity. This division, under Colonel Sherbrooke, was accompanied by General Baird, and reached the eastern face of the fort in less than an hour, without having met any serious opposition except near the Mysore Gate, where many men were killed and wounded, and in spite of meeting with less resistance than the left attack, their casualties somewhat exceeded them.

Tippoo was killed during the assault. "Nine hundred and twenty-nine pieces of ordnance were found in the fort, of which two hundred and eighty-seven were mounted on the works. There was also a very large quantity of gunpowder, round shot, small arms, and military stores of different kinds. The artillery, however, when examined in detail, does not appear to have been of a very formidable description, as there were no fewer than 436 guns throwing balls under five pounds. Out of 373 brass guns, 202 were from Tippoo's own foundry, 77 were English, and the rest French, Dutch, and Spanish; of the 466 iron guns, only 6 were from Tippoo's foundry, 260 having been of foreign, and 200 of English make; of 60 mortars and cohorns, 22 were Tippoo's, the rest English and foreign. The howitzers, 11 in number, had, with one exception, been cast in Seringapatam."

The 2nd Battalion, 3rd Regiment, between the 4th April and 4th May 1799, lost 1 European killed and 2 wounded, 12 natives killed, 47 wounded, and 15 missing.

Copy of a General Order by the Commander-in-Chief:—

"CAMP AT SERINGAPATAM, *5th May* 1799.

"The Commander-in-Chief congratulates the gallant army which he has the honour to command, on the conquest of yesterday. The effects arising from the attainment of such an acquisition as far exceed the present limits of detail, as the unremitting zeal, labour, and unparalleled valour of the troops surpass his power of praise. For services so incalculable in their consequences, he must consider the army as well entitled to the applause and gratitude of their country at large."

Copy of a General Order by Government:—

"FORT ST. GEORGE, *15th May* 1799.

"The Right Honourable the Governor-General in Council, having this day received from the Commander-in-Chief of the allied army in the field, the official details of the glorious and decisive victory obtained at Seringapatam on the 4th May, offers his cordial thanks and sincere congratulations to the Commander-in-Chief, and to all the officers and men composing the gallant army which achieved the capture of the capital of Mysore on that memorable day. . . .

"Under the favour of Providence, and the justice of our cause, the established character of the army had inspired an early confidence, that the war in which we were engaged would be brought to a speedy, prosperous, and honourable issue. But the events of the 4th May, while they have surpassed even the sanguine expectations of the Governor-General in Council, have raised the reputation of the British arms in India to a degree of splendour and glory unrivalled in the military history of this quarter of the globe, and seldom approached in any part of the world. The lustre of this victory can be equalled only by the substantial advantages which it promises to establish, by restoring the peace and safety of the British possessions in India on a durable foundation of genuine security.

"The Governor-General in Council reflects with pride, satisfaction, and gratitude, that in this arduous crisis the spirit and exertion of our Indian army have kept pace with those of our countrymen at home; and that in India, as in Europe, Great Britain has found, in the malevolent designs of her enemies, an increasing source of her own prosperity, fame, and power."

After the siege the regiment formed part of the Seringapatam garrison, while General Harris encamped in the neighbourhood with the main body of the army to make arrangements for the settlement of the Mysore country.

On the opposite page is given a list of officers of the 3rd Regiment "agreeably to the new arrangements establishing the regimental rise on the Madras establishment," dated 22nd January 1800.

Whilst in garrison at Seringapatam the regiment furnished five companies for field service, with the detachment under Lieutenant-Colonel Tolfrey, against Kistnapah Naik, the Rajah of Bullum, who had taken possession of the ghaut leading from Mysore into Canara, thus interrupting communication with Mangalore. Colonel Tolfrey's force, consisting of 13 companies of native infantry, 2 guns, 50 pioneers, and a body of Mysore Horse, arrived at Eygoor on the 30th March 1800, and, finding it abandoned, he destroyed the place, and advanced to Arrakaira near Munzerabad, where the Rajah occupied a strong stockaded position in thick forest. The position was attacked on the 2nd April, but the British were repulsed with the loss of 47 men killed and wounded. About the end of the month, a reinforcement of British and native troops, under Colonel Montressor of H.M.'s 77th Regiment, having arrived, the place was carried by storm on the 30th April 1800. In this affair the regiment lost 2 killed and 12 wounded.

1800

The following is a copy of the General Order by

LIST OF OFFICERS OF THE 3RD REGIMENT "AGREEABLY TO THE NEW ARRANGEMENTS ESTABLISHING THE REGIMENTAL RISE ON THE MADRAS ESTABLISHMENT."

(From the Manuscript Records in the India Office, dated 22nd January 1800.)

Names.	Regimental Rank.	Battalion.	Remarks.
Colonel—			
Edward Collins . .	3rd October 1792		
Lieutenant-Colonels—			
Barry Close . . .	29th November 1797	1st	Resident at Mysore, and Hon. A.D.C. to the Gov.-Gen.
John Chalmers . .	31st July 1799	2nd	
Majors—			
John Bannerman . .	29th September 1798	1st	
John Kennet . .	31st July 1799	2nd	
Captains—			
William Sheppard .	8th January 1796	1st	
Alexander Allan . .	8th January 1796	2nd	
Poole H. Vesey . .	29th November 1797	1st	
Ronliston Robinett .	12th October 1798	2nd	
Charles Trotter . .	26th December 1798	1st	
Charles Burton Phillipson	31st July 1799	2nd	
Thomas Boles . .	10th December 1799	2nd	
Captain-Lieutenant—			
James Welsh . .	10th December 1799	1st	Adjutant & Quartermaster
Lieutenants—			
John Lloyd-Jones .	6th March 1793	2nd	
Charles Aldridge . .	11th December 1793	1st	
Charles Lucas . .	6th August 1794	2nd	Adjutant 2nd Batt.
George Pippard . .	6th August 1794	2nd	
Michael Egan . .	14th July 1795	1st	
Thomas Little . .	8th January 1796	1st	Adjutant 1st Batt.
Henry Davy . .	8th January 1796	1st	
Francis Evans . .	4th March 1797	2nd	
Joseph Knowles . .	20th August 1797	1st	
Charles Lester . .	29th November 1797	2nd	
John Rendall . .	2nd August 1798	1st	
George Wilson . .	12th October 1798	2nd	
Hastings M. Kelly .	24th October 1798	2nd	
Hercules H. Pepper .	26th December 1798	1st	
John Carfrae . .	26th December 1798	2nd	
Hugh Dalrymple . .	...	1st	
Thomas S. Stevenson .	...	2nd	
Surgeon—	...	...	
Assistant Surgeons—	...	...	

Major-General Braithwaite, dated Choultry Plain, 8th May 1800:—

"Major-General Braithwaite has received from the Honourable Colonel Wellesley a report of an attack on the Barriers of the Bullum Rajah at Arakerry, made on the 30th ultimo by a detachment under the command of Lieutenant-Colonel Montressor, of H.M.'s 77th Regiment, in which, after an obstinate resistance, the Post was completely forced.

"Major-General Braithwaite considers Lieutenant-Colonel Montressor as entitled to his particular thanks for the able manner in which he directed the efforts of the detachment under his command; and he cannot more properly express his approbation of the conduct of the Other Officers and Troops employed, than by publishing to the Army the following extract from Lieutenant-Colonel Montressor's report to the Honourable Colonel Wellesley on this occasion":

EXTRACT.

"I did myself the honour to write to you on the 28th inst., stating it was my intention to march to Munzerabad, and attack Kistnapah Naig at Arakerry. I accordingly marched to Munzerabad on the 29th, and on the following morning (this day), after leaving my equipage and stores under the protection of the guns of that Fort, and of the Rajah of Mysore's Cavalry, I attacked and carried Arakerry, disposed the Poligar's adherents, and burnt several of his villages and magazines of grain. I am much indebted to the Troops under my command for the zeal and gallantry displayed throughout the day.

"The column of the attack consisting of the flank companies of H.M.'s 73rd and 77th Regiments under Captain M'Pherson, three companies of the 2nd Battalion, 3rd Regiment, and grenadiers of the 1st Battalion, 12th Regiment, was led by Major Capper with a degree of spirit and gallantry which overcame a continued range of obstacles and resistance for near a mile and a half through a most intricate country. The Reserve, under Lieutenant-Colonel Tolfrey, was conducted with considerable judgment. I am also indebted to Captain Colebrooke of the Guides for volunteering his services in the line; and I trust I may be permitted to express a sentiment of gratitude and regret when I mention that gallant and meritorious officer Captain Grose of the Pioneers, who unfortunately was killed early in the day, in endeavouring to place some ladders at one of the barriers."

The following is a copy of the General Order by Government, dated Fort St. George, 12th May 1800:—

"The Right Honourable the Governor in Council having received from the Honourable Colonel Wellesley a report of the successful attack of the post of Arakerry by a detachment of H.M.'s and the Honourable Company's Troops under Lieutenant-Colonel Montressor, the details of which have been published in General Orders by the Officer Commanding the Army in Chief; his Lordship deems it due to the Officers and Men of that detachment to express in public orders his approbation of their conduct as stated in the report of the Officer Commanding, and his Lordship has great satisfaction in observing that the judicious disposition and spirited direction of the attack was entirely worthy of the distinguished Military Character of Lieutenant-Colonel Montressor.

"By order of the Right Honourable the Governor in Council,

(Signed) "G. BUCHAN, Sub-Secretary."

The detachment left Bullum in June, and the five companies of the regiment presumably rejoined regimental headquarters at Seringapatam, for there is no record of any portion of the regiment having taken part in the fighting in the Wynaad and Carnatic during the years 1801 and 1802.

The following is a copy of a General Order by Government, dated Fort St. George, 3rd July 1800:—

"The Right Honourable the Governor in Council has been pleased to direct, at the recommendation of the Officer Commanding the Army in Chief, that in consideration of his long, active and zealous services, Subidar SELAZEE ROW of the 2nd Battalion, 3rd Regiment, of Native Infantry, shall be pensioned on his full pay, and receive a monthly allowance for a Palanquin of Ten (10) Pagodas from the first of the present month (July) . . . and also that the individuals mentioned in the following list, who have been transferred to the Pension List, shall receive the net infantry pay of their respective ranks:—

"2nd Battalion, 3rd Regiment:—Havildars Rayman Sing and Narupah, Naigue Condiah, Sepoys Shaik Ismail, Mootoo Naig, and Sied Cumaul."

1803

War having been declared with the Mahrattas in 1803, the 2nd Battalion, 3rd Regiment, under the command of Major Kennet, received instructions to hold itself in readiness for active service. On the 8th February, a force, consisting of 1 European regiment, 1 native cavalry regiment, 6 battalions of native infantry (inclusive of the 2nd Battalion, 3rd Regiment), 2000 of Poorniah's horse, and 5000 of his infantry, made its first movement from Seringapatam for the Mahratta country. After a very hot and uninteresting march of two hundred miles, the force arrived near the Grand Army under General Stewart, and the Centre Army under General Campbell, on the plains near Hurryhur on the 8th March. On arrival, orders were issued for the army to march at once. The infantry was formed into two brigades. The whole force now numbered 10,617 men;

and the 2nd Battalion, 3rd Regiment, 998 strong, in company with the Scotch Brigade, the 1st Battalion, 2nd Regiment, and the 2nd Battalion, 12th Regiment, formed the 1st Infantry Brigade, under Lieutenant-Colonel Harness of H.M.'s 80th Foot. Major-General Wellesley assumed command of the Field Army, and on the 9th March the army left Hurryhur, and began its march to Poonah, which had been seized from the Peshwar by Holkar in October of the previous year. Holkar had, however, left his brother Amrut Row in charge, with orders to burn the city on the approach of the British. The General, having heard of this, made a forced march of forty-two miles on the night of the 19th March with the cavalry and 2nd Battalion, 12th Regiment, and arrived at Poonah on the following morning, in time to save it from the flames. The remainder of the army arrived on the 22nd, and in this neighbourhood the troops remained in camp and inactive for six weeks; the army not moving ground till the 4th June. In the interim, however, the Peshwar came back to his capital, and the "2nd Battalion, 3rd Regiment, was exchanged for H.M.'s 78th Regiment"; the Scotch Brigade having already been sent to join Colonel Stevenson's force, at some distance off. On the 4th June, the General marched towards the Godavery to watch the armies of Scindia and the Rajah of Berar, who were suspected of intending to invade the territory of the Nizam. The 2nd Battalion, 3rd Regiment, remained behind to garrison Poonah. "Very much in the dark with regard to Indian politics, we had naturally concluded, that as we came to succour the Peshwar, his friends would be our friends, and his foes our likeliest opponents; but here we reckoned without our host, for the man we were now to attack was not Holkar, who had deposed him, but Scindia, who had upheld him, and actually suffered a defeat, near Poonah, in his cause!" Ahmednuggur was taken by Wellesley on the 4th August 1803, and the 2nd Battalion, 3rd Regiment, under the command of Captain Lucas (the Commandant, Major Kennet, having lately died), moved up from Poonah to garrison the fort, whilst Wellesley moved northwards in pursuit of the enemy. Captain Lucas was appointed Commandant of the fort, and Captain Carfrae of the same regiment, paymaster.

Colonel Welsh in his *Reminiscences* gives the following account of the action at Korjet Corygaum in September 1803:—

"A company of the 12th Regiment of Native Infantry, under Lieutenant Morgan, having been detached from camp to proceed to the Carnatic, in charge of various drafts from native corps in our army, for new corps raising at Madras; along with this party, and taking advantage of their escort, were Captain O'Donnell and Lieutenant Bryant of the 2nd Native Cavalry, proceeding to join their corps with the force under General Campbell. They had reached the vicinity of a village called Kurjet Koriagaum, about seventy miles from Ahmednugger, when they were suddenly attacked by a body of about one thousand five hundred men, the former garrison of Ahmednugger, of whom at least one-third were Arabs. Captain O'Donnell, who, though small, was a truly gallant fellow, immediately assumed the command, and led on his motley band, amounting, on the whole, to not more than one hundred men, to the charge. Lieutenant Bryant, a very powerful man, first saved the life of O'Donnell, who had snapped his pistol at the leader of the Arabs, and was about to be cut down by him, when Bryant put him to death; and then attacking their colour-bearer, cut him down also, and seized their standard. At this moment the enemy's cavalry appeared, and Captain O'Donnell drew off his little party into the village; but so closely were they pursued, that they were forced to take post in a large choultry, from whence the enemy could not dislodge them. Here the extraordinary strength and courage of Bryant, if it did not entirely save their lives, at least conduced to their preservation from famine. He harangued the sepoys in broken English, not knowing a word of any native language, and continually sallied out with a few volunteers, in search of food, and as regularly killed some of their opponents. . . . Of the hundred men collected and blockaded in this spot, all the native officers behaved ill, and would have persuaded the men to capitulate, had not many of them taken courage by the behaviour of Lieutenant Bryant, to them a perfect stranger, and by the conduct of the other two European officers. . . . Matters were in this state, when Captain Lucas, with four companies of the 2nd Battalion, 3rd Regiment, and two guns, made his appearance, and relieved them, without striking a blow, for the enemy had withdrawn, aware of his approach; and, acting strictly up to the orders he had received, 'to make no delay, and risque nothing beyond the relief of the party,' he would not attack their camp outside the village, nor suffer any of his detachment to meddle with them; but marched back as fast as he came, and enabled us to move off to join the army, the party thus relieved returning with us. As it is always easy to find fault, Captain Lucas was very generally blamed for not attacking the Arab camp, only two or three miles out of his way, when the very well-being of an army depended on his security and speedy return. In my mind he acted as became a soldier. I do not know the casualties of the little party, but believe they were numerous the first day, and that they lost their horses and all their baggage during their retreat into the village."

On the 31st October 1803, a convoy for 1500 bullocks (carrying grain for the army) composed of three companies of the 2nd Battalion, 3rd Regiment, two of the 2nd Battalion, 10th Regiment, and 400 Mysore Horse, all under Captain Baynes, was attacked by 5000 horse belonging to the Rajah of Berar at the village of Umber, half-way between the Godavery and Jaulna, who were repulsed with considerable loss, particularly in horses. The convoy joined the General at Aurungabad the next day. Captain Baynes was thanked for the able disposition he had made of his small force, and the steadiness of the officers and men was favourably noticed in the same Order.

Towards the end of 1803, Scindia and the Rajah of Berar sued for peace, and treaties were concluded with them, which the Governor-General in Council ratified in January and February 1804. Soon after this the Madras troops were ordered to be withdrawn to their own Presidency, and the relief commenced about the middle of the year, but, owing to the threatening attitude of Scindia, three battalions stood fast till 1806. Of these the 2nd Battalion, 3rd Regiment remained at Ahmednuggur until March of that year, when, a fresh treaty with Scindia and peace with Holkar having been concluded, the regiment returned south and was quartered at Trichinopoly.

1804 In General Orders of the 10th November 1804, the resolutions of the House of Commons were published to the army, of which the following is an extract:—

"JOVIS, *3rd die Mai*, 1804.

"Resolved *nemine contradicenti* that the thanks of this House be given to Major-General the Honourable Arthur Wellesley for the many important, brilliant, and memorable services achieved by him in command of the separate army within the Dekkan, and also to the several officers of the army, both European and native, for their gallant conduct and meritorious exertions during the arduous, honourable, and successful campaign in the East Indies.

"Resolved *nemine contradicenti* that this House doth highly approve and acknowledge the zeal, discipline, and bravery uniformly displayed by the non-commissioned officers and private soldiers, both European and native, employed against the enemy in the East Indies; and that the same be signified to them by the commanders of the several corps, who are desired to thank them for their exemplary and gallant behaviour. . . .

(Signed) "J. LEY, Cl. D. Dom. Com."

On the 27th March 1805, the men of the battalion were ordered to be instructed "in the duties of light troops as recently introduced in Europe. The instruction so imparted was not to be limited to the light companies, but every company was to be taught to manœuvre in extended order." 1805

The 2nd Battalion, 3rd Regiment, appears to have remained in garrison at Trichinopoly until the end of 1808, when, complications having arisen in Travancore, the regiment joined the force under Colonel the Honourable A. Sentleger, which was assembling with the object of operating in Travancore from the south, to create a diversion in favour of Colonel Chalmers's column, which was threatened near Quilon by some 30,000 of the enemy with 18 guns. 1808

"The southern fortified lines of Travancore, commencing among rugged hills on the seacoast, near Cape Comorin, were carried on, joining such hills as came in the way, as far as the mountainous range which separates the eastern from the western coast; these fortifications completing the boundary of that country. They were divided into two separate parts by a high mountain, those next the sea being called the 'Southern Lines,' and those carried beyond that mountain, to the ghauts, the 'Arambooly Lines.' It was against the latter that our operations were intended, because the high road from Palamcottah passed through the centre of them, by a gate covered with two large circular bastions, and defended by several pieces of ordnance. The extent of the whole might be about two miles, embracing a rugged hill to the southward, completely fortified, and a very strong rock, about half-way, called the 'Northern Redoubt,' beyond which, to the range of mountains, it was nearly inaccessible in deep jungle, The works consisted of small well-built bastions for two and three guns, the whole cannon proof and covered by a thick thorny hedge, the approach to which was rather difficult, from the wild state of the country, within cannon range of the walls." Before Sentleger's force could enter Travancore, it was thus necessary to force the Arambooly Lines. On the 6th February he marched six miles, and took up a position within five miles of the works, the 2nd Battalion, 3rd Regiment, under Captain Lucas, being detached in advance the same evening.

Sentleger, being without a battering train, determined to attempt a surprise, and on the night of the 9th February 1809, twelve companies of infantry, of which seven, under the command of Captain Lucas, belonged to the 2nd Battalion, 3rd Regiment, and a party of pioneers with scaling ladders—the whole under Major Welsh of the 1st Battalion, 3rd Regiment,—moved towards the redoubt on the hill at the southern extremity of the works. The country being very difficult, owing to the thickness of the jungle and the numerous ascents and descents, the troops did not reach the foot of the walls near the summit of the hill until after a march of six hours, when the ladders were planted, and the whole of the works, together with the arsenal, were in our possession by eight o'clock on the morning of the 10th. The British loss was extremely small, viz. 1 European and 1 sepoy killed, 4 Europeans and 7 natives wounded.

1809

The following officers belonging to the 3rd Regiment

were present with the detachment during the escalade under Major Welsh :—Captain Lucas, Captain Pepper, Captain Carfrae, Lieutenants Walker, Tagg, Dawson, Goble, Inverarity, Jeffrey, Rule, Shepherd, and Blake.

The following is an extract from an Order published by Lieutenant-Colonel Sentleger in consequence of this capture :—

"Lieutenant-Colonel the Honourable A. St. Leger has much satisfaction in conveying to the troops under his command, the most sincere congratulations on the brilliant achievement this morning. . . . The Lieutenant-Colonel requests Major Welsh will convey to the officers and men who composed the detachment for escalade, under his command, the most unqualified approbation of their gallant exertions in accomplishing an object which must ever be considered as entitled to a high place in military records."

Lieutenant-Colonel Sentleger describes Major Welsh's success in the following quaint language :—

"The southern redoubt, which presented a complete enfilade of the whole of the main lines as far as the gate, was the object of Major Welsh's enterprise ; an undertaking which, from the natural strength of the approach, appeared to be only practicable to the exertions and determined bravery of British troops, led on to glory by Major Welsh. It was ascended under cover of the night, and our troops had actually escaladed the wall ere their approach was suspected ; and the ascent was of such great difficulty as to require six hours of actual scrambling so as to reach the foot of the walls. In consideration of the brilliancy of the achievement, I feel a pleasurable duty in detailing, for the information of the Honourable Governor in Council, the names of those officers who accompanied the detachment for escalade, which consisted of two companies, and the picquet of H.M.'s 69th Regiment, commanded by Captain Syms ; the four flank companies and five battalion companies of the 3rd Regiment, under Captain Lucas ; and it did not require that confirmation which Major Welsh has conveyed, in the most handsome manner, to convince me that to have accomplished such an object, every man must have done his duty. . . . When Major Welsh had once effected his security in this commanding position, I despatched to his assistance, by the same arduous route, a company of H.M.'s 69th and three companies of the 1st and 2nd Battalions of the 13th Regiment to reinforce and add confidence to his party. As soon as this addition was perceived, a detachment from his party stormed the main lines, and by dint of persevering bravery, carried them entirely, when the northern redoubt was abandoned by the panic-struck enemy, who fled in all possible confusion in every direction, leaving me in possession of their strongest lines."

The following complimentary Order in connection with this was issued on the 16th February 1809 :—

"The Governor in Council considers the execution of the above service to reflect equal honour on the judgment with which it was planned, and on the signal zeal and bravery with which it was carried into effect. The Governor in Council accordingly conveys to the Honourable Lieutenant-Colonel Sentleger the expression of his warmest approbation and thanks, and Lieutenant-Colonel Sentleger is requested to make known to the officers and troops under his command, particularly to Major Welsh, who gallantly and ably conducted the party employed in the assault, and to the other officers mentioned by Lieutenant-Colonel Sentleger, the sentiments of cordial approbation due to their meritorious conduct."

On the capture of the Arambooly Lines on the 10th February 1809, the army encamped the same day inside the walls in a convenient position, two miles from the Arambooly Gate. The arsenal, which was well filled with arms, ammunition, and many stores, with a quantity of valuable ordnance in the works, was taken possession of. Parties were at once detailed to destroy the works on both sides of the gate, which was, however, left entire, to secure communications. Prize money was afterwards divided amongst the troops who had taken part in the capture.

On the 17th February the army marched for the interior; the advance guard, commanded by Lieutenant-Colonel M'Leod of the 69th, consisted of the flanking companies of the 69th Regiment, 350 Kaffirs (who had just joined the army from Ceylon), and six native flank companies (two of which were furnished by the 2nd Battalion, 3rd Regiment), and the cavalry with six guns. "This party moved off from the right of the line at three o'clock a.m.; the line following at half-past four, and thus leaving a distance of three miles between them. Having got on six miles by daybreak, they found the enemy strongly posted in a village, across a river with high banks, commanding the approach, and several cannon pointed down the high road. Their force was supposed to amount to six hundred men, and they had every advantage, in point of position, that men could desire. Colonel M'Leod immediately formed his line for attack, and drove the enemy from their guns, after a very heavy fire of both cannon and musketry, which, unfortunately, did considerable execution, owing to the exposed situation of our troops in advancing. The enemy were completely routed, and dispersed in all directions for some miles. The country was said to be too difficult for the cavalry to follow them, which doubtless saved many, if not the whole, from total destruction. Lieutenant Charles Johnstone, however, with a small party of horse, contrived to get in amongst them, and did some execution. Nine capital guns, and several dead bodies, were the fruits of this victory; in addition to which, we gained possession of two very fine villages, called Cotour and Nagercoil." The British losses amounted to two officers and forty-nine rank and file wounded, of whom, however, Colonel Sentleger reported that "by far the greater part are only slightly wounded."

After this affair, the army encamped four miles beyond the village of Nagercoil, which was distant about eight miles from Arambooly.

"Although, generally speaking, the enemy had proved far below our expectations, yet there were some exceptions . . . these instances, however, were rare; for, taking them all in all, I never beheld a more dastardly crew; nor did they deserve the name of soldiers, although neatly clothed in military uniforms, furnished with capital arms, and in a country every inch of which might have been defended."

On the 19th February, Major Welsh "had the honour to lead the advance, consisting of the picquets, and some flank companies with two 6-pounders, expecting hard work, though the line was not very distant in our rear. After proceeding three or four miles, we met some peaceable villagers, who informed us that the two forts of Oodagherry and Palpanaveram in our front had been abandoned by the enemy, which was the first time we had heard of such fortifications, though we had been expecting to find some field-works to be taken. The news soon spread, and ere we had advanced much farther, we could distinguish white flags flying on trees and sticks, when the whole headquarter gentlemen passed us, preceded by some troops to explore the way. Shortly afterwards, the road led us on a sudden within musket shot of stone bastion and curtain, mounting several cannon pointed on the road, and we found this to be part of the fort of Oodagherry, with white flags flying, and not a soul within. I was directed to take possession of both forts with my own corps; and Palpanaveram being the largest, a mile farther on, I left two companies in Oodagherry, and, proceeding thither, disposed of the corps in an open space in the centre, posting Hindoo guards in all the pagodas, and the officers taking possession of a large and very well-built palace belonging to the Rajah. Here we found many valuable swords, dirks, pistols, guns, spears, rich muslins, kincobs, etc., as well as thousands of jewel boxes, broken open and pillaged by the flying enemy, to give us some idea of what we had lost. Several of the swords proved to be gold-hilted, and the blades were of the first water. Of course all we could lay hands on were secured as prize property, and afterwards sold by public outcry.

"Whilst we thus were advancing from the southward, the subsidiary force at Quilon was by no means idle. Shut up in the heart of a strong country, with the inhabitants all in arms against them, they had several severe actions, but invariably came off conquerors. Nevertheless their situation was daily becoming more critical, until the news of our entering the lines reached the masses by which they were surrounded, when, giving up every hope of further success, they dispersed and fled in all directions; for those lines, ill as they were calculated to resist an English force, had hitherto been deemed impregnable; and Tippoo, in the zenith of his power, had been repulsed from them with great loss."

On the 27th February, Sentleger commenced the march to Trevandrum, the capital of the Rajah, but did little more than break ground on that day. "On the 28th the troops moved by the right, and the baggage on the left, with orders, in case of an attack, as the Rajah had disowned his ministers' acts (an armistice having been proclaimed in our force on the 26th February), that each corps should countermarch on its own ground, a thing totally impossible in such a country as we had to pass through. My reason for inserting this strange march is, that it was something out of the ordinary course of modern tactics; and as we had no enemy to oppose, it served to amuse and beguile the tedious hours. Several of our guns and limbers were upset on the road . . . but at length we encamped, on the night of the 28th, upon high and very uneven ground covered with bushes, in the most perfect disorder imaginable.

"On the 2nd of March we reached the neighbourhood of the capital, and encamped, as well as we could in so intricate a country, near a fine deep river, with a good bridge across, about three miles to the southward of the town."

Colonel Chalmers's force having by this time arrived a few miles north of Trevandrum, the Rajah quietly submitted, and it was agreed that, in addition to all arrears of subsidy, he should pay the expenses of the war, and that the Carnatic Brigade and the Nair Battalions in his service should be dismissed.

"March the 15th, the new Dewann paid a formal visit to our camp, where he was received with military honours, and a salute of fifteen guns. All the native officers of His Highness's late Carnatic Brigade being brought into the camp prisoners, and disgraced by the drummers of the line, who cut their jackets off their backs, and then turned them out, with the 'Rogue's March.'"

About the 20th March, the force returned by corps to Oodagherry, to be cantoned there, the last arriving on the 8th April.

The force under Colonel Sentleger again received, from the Governor in Council, the "warmest thanks for the activity, zeal, and bravery which have signalised their operations," in General Orders of the 26th February 1809.

Shortly after this settlement of affairs in Travancore, the battalion marched for Mysore, and relieved the 1st Battalion, 8th Regiment, at Bednore (furnishing a detachment at Bangalore), where, on the 9th August 1809, Major Lucas the Commandant, with the rest of the British officers, refusing to sign a memorandum of confidence in and obedience to the Governor of Fort St. George, Sir George Barlow, made over charge of the battalion to the senior native officer, pending the nomination of another European Commandant.

Lieutenant-Colonel Orrok was soon after this appointed to the command, and Lieutenant Cadenskie of H.M.'s 80th Regiment to the acting Adjutancy of the battalion. The regiment garrisoned Nundydroog for some months, moving in 1810 to Seringapatam, and the same year marched into quarters at Cannanore. Early in 1810, the Supreme Government having resolved to attempt the complete reduction of the French islands, Bourbon and Mauritius, the 2nd Battalion, 3rd Regiment, immediately they heard of the contemplated expedition, volunteered for service. Major-General Gowdie, who was Commanding-in-Chief at the time, in forwarding the application, with some others, for the consideration of the Governor-General, remarked—

1810

"The subject of these reports contains so unequivocal a proof of the fidelity, zeal, and attachment of the native army, that Major-General Gowdie cannot suppress the expressions of his earnest hope that they may meet the public notice of Government, nor are the merits of the European officers, who have so conspicuously evinced their zeal for the public service, less entitled to the Major-General's approbation, and he has desired me to request that in bringing these reports to the notice of Government you will do him the favour to notice the following officers in command of corps who have distinguished themselves by a zeal for the public service, and an example of the fidelity and worth of the native troops, as honourable to themselves as it must be gratifying to Government.

. . . .

"Lieutenant-Colonel Orrok commanding 2nd Battalion 3rd Regiment.'

. . . .

On the 6th April, the following General Order was published with reference to the foregoing extract :—

"The Right Honourable the Governor-General in Council is happy to observe that the confidence and discipline of the native army in this Presidency, which was manifested by the recent determination of the Government to employ a respectable detachment from it upon foreign service, has been fully justified by the alacrity and ardour with which not only the troops selected for that duty, but several other corps, have volunteered for foreign service, and his Lordship requests that Major-General Gowdie will convey to the whole of these troops the expression of his warm approbation of the zeal, fidelity, and military spirit by which their conduct has been distinguished on this important occasion. The Right Honourable the Governor in Council is happy to avail himself of this opportunity of expressing his fullest approbation of the meritorious and zealous exertions, on this occasion, of the officers whose names have been brought to his notice by Major-General Gowdie."

On the 2nd November 1810, Government announced that a reward of five hundred pagodas would be given to every young officer who, on due examination, "should be found to have made himself master of the Hindostanee language."

The medals which had been struck in honour of the capture of Seringapatam, only arrived in Madras in January 1811, although the following extracts, from a letter from the Court of Directors regarding them, had been published to the army as far back as the 6th July 1808 :—

1811

"Some time ago we caused a medal to be executed by one of the most eminent artists in this country, in commemoration of the brilliant success of the British arms in Mysore in 1799, for distribution amongst the officers and soldiers (European and native) employed on that glorious occasion.

On one side of it is represented the storming the breach of Seringapatam, from an actual drawing on the spot, with the meridian sun denoting the time of the storm, with the following inscription in Persian underneath: 'The Fort of Seringapatam, the Gift of God, 4th May 1799.' On the reverse side is the British Lion subduing the Tiger, the emblem of the late Tippoo Sultan's government, with the period when it was effected, and the following words in Arabic on the banner: 'Assad-oollah-ul-Ghalib,' signifying the Conquering Lion of God."

"Of these medals, *gold* ones have been struck for His Majesty, the Right Honourable Lord Melville, the Governors in India at the time, Marquis Cornwallis, the Nizam and his two Ministers, the Peshwa and his Minister, the Nabobs of Arcot and Oude, and the Rajahs of Tanjore, Travancore, Mysore, Coorg, and Berar, Dowlut Row Scindiah, the Commander-in-Chief, General Officers on the Staff employed on the service, and the Oriental Museum. *Silver gilt* for the Members of Council at the three Presidencies, the Residents of Hyderabad and Poonah, the Field Officers, and General Staff on the service. *Silver* for the captains and subalterns on the service. *Copper bronze* for the non-commissioned, and pure *graven tin* for the privates."

On the arrival of these medals, orders were issued for their distribution to survivors, whether effective or otherwise, and also to the heirs of deceased persons who had been entitled.

"In addition to such customary methods of securing the allegiance of its native soldiers as pay, family remittances, and pensions, Government, in days gone by, frequently rewarded such of them as had done well in what may be called an informal, but which was certainly a most popular and much appreciated way. It made them presents of money, a sword, or a special medal struck for the occasion; it gave them lands or the honours accruing from the possession of a palanquin; and when they were dead, it sometimes perpetuated its sense of the services they had rendered whilst alive by according a grant for the maintenance and repair of their tombs, so that all might see that the Sirkar did not forget departed worth, or the brave soldiers who had freely offered their lives in its service."

The following copy of G.O.G. of the 12th November 1811, gazetting a reward of this nature, is interesting:—

"In consideration of the long and faithful services of the under-mentioned officers, the Honourable the Governor in Council has been pleased to direct that a palanquin shall be presented to each of them with the usual allowance of twenty pagodas per month for the support of that equipage ; and, further, that on their demise respectively, the nearest heir of each of these officers shall be pensioned on the half-pay of a subadar of cavalry or infantry, according to the branch of the service to which each individual belongs, viz.—

. . . .

"Sheick Esmael, 2nd Battalion, 3rd Regiment, Native Infantry."

. . . .

The troops stationed in Cannanore in 1811 were H.M.'s 30th Regiment, with the 2nd Battalion of the 3rd, 2nd Battalion of the 9th, and 1st Battalion of the 15th. The 2nd Battalion, 3rd Regiment, furnished a detachment of two companies, under the command of Captain James Tagg, at Manantoddy in the Wynaad.

In this year, Colonel Webber succeeded Lieutenant-Colonel Orrok in the command of the regiment. The inhabitants of the Wynaad had always been turbulent, and a force under Colonel Stevenson had been employed for many months in quelling a rebellion which had broken out, when, being lulled into a mistaken sense of security, all the troops, excepting the two companies at the headquarter station commanded by Captain Tagg, had been withdrawn. In March 1812, a rebellion broke out at Manantoddy and Sultan's Battery, which latter place was now occupied by thirty men of the regiment, under a native officer. Both these places were threatened by the insurgents. The post of Manantoddy had been well chosen, being on the top of a moderate-sized hill, clear of wood, and commanding the high road on both sides, and it had never been deemed necessary to fortify it thoroughly. Previous to the breaking out of this rebellion, Captain Tagg had not observed any particular indications of discontent, and was therefore without any supplies on the hill ; a good bazaar, and all the native huts, being situated at the foot of it. There was also another thickly wooded hill on the other side, completely commanding this bazaar, which, he was one

1812

morning informed, had been plundered; and, on going towards the corner of the hill, he was shot at from the opposite side. He immediately collected a small party, and, dashing down, recovered as much grain, etc., as he could, and sent it up the hill with all the sepoys' families, who were in too exposed a situation below. Returning to his post, he at once sent off the intelligence to Mysore and Cannanore, and prepared for resistance. The supplies they had served for a few days, when, running short, he had made up his mind to force the enemy's nearest post on the neighbouring hill, and to follow them to some place where he might find a store of grain. This was, however, rendered unnecessary by the timely arrival of a detachment from Cannanore, consisting of one company of H.M.'s 30th Regiment, the 2nd Battalion, 3rd Regiment, and a detachment of the 1st Battalion, 5th Regiment, all under the command of Lieutenant-Colonel Webber, 2nd Battalion, 3rd Regiment. A force from Seringapatam, consisting of one company of H.M.'s 80th Regiment, two companies 1st Battalion, 3rd Regiment, and two companies 1st Battalion, 13th Regiment, under the command of Major Welsh, had also been ordered to march to Captain Tagg's relief and put down the rebellion. Colonel Webber's column was attacked whilst ascending the Cotiaddy Pass, and Captain Hunter, Lieutenant Inverarity, and seventeen or eighteen men of the 2nd Battalion, 3rd Regiment, were wounded, the two former severely. The rebels, however, dispersed, and on the 9th April 1812 the post at Manantoddy was relieved.

Major Welsh arrived next day, "after a very tedious and laborious march, in which the line was suddenly assailed by a flight of arrows from both sides of the road, by which two soldiers and one sepoy were wounded, and an English dog killed, with a part of our force only; and such was the thickness of the jungle, that I was totally ignorant how the rear were coming up. Applying, therefore, to Colonel Webber for some fresh men and officers, I returned with Captain Pepper with this reinforcement, and reached our rear-guard, which had taken post six miles off, at midnight, where we remained with them till daylight, suffering much from cold, hunger, and thirst, not being able to get even a little water all night. As the day broke, we found ourselves entirely masters of the field, with broken and upset carts and baggage strewed in every direction. Some hours of rest, though not of sleep, had prepared our men for fresh exertions, and all was snug at Manantoddy in the course of a few hours."

Having all united under Colonel Webber's general command, Himinutear Cawn, the Mysore general, was requested to bring up the supplies he had collected on the frontier, and a plan of operations decided on.

On the 14th April, three parties were formed to scour the jungle in different directions; one under Colonel Webber, a second under Major Welsh, and a third under Captain Jones. After an early breakfast, all three parties set out to the northward in search of the enemy, but could not fall in with them, and after making a very tedious circuit in deep jungle, and destroying the houses of several chiefs, met on a small hill called Trichilary, near a celebrated pagoda. In this day's work twenty-two miles were covered.

On the 15th, two parties were formed, but no more rebels in arms were seen. Many of them, however, came in to the Collector, and pledged themselves to give up their leaders in six days, on a promise of pardon to the rest.

"On the 16th April, Colonel Webber allowed me (Major Welsh) to choose a party of officers and sepoys, without Europeans, who, deprived of their little comforts, could not stand the fatigue and exposure, and to try what I could do in the hunting way. Leaving, therefore, our artillery, Europeans, and three of our flank companies at Manantoddy with the Colonel's party, and arming each officer, European and native, with an artillery fusil and bayonet, we set out, the officers being Captain Pepper, Lieutenants Tagg, Williamson, and Meredith, with two hundred sepoys from both battalions of the 3rd, and a respectable native, who knew the country. As we set off in the evening, and had neither camp equipage nor baggage, beyond absolute necessaries, we took post that night on a high hill about eight miles and three quarters south from Orekody, called Coominah, all the rest of the force returning at the same time to Manantoddy. On the 17th we set forward at nine o'clock a.m., after a very heavy dew, which wet everything through, and rendered us all most uncomfortable; no tidings of the rebels yet; and being in the evening close to a deep river, said to be full of alligators, we pitched our only little tent on the bank, when a heavy fall of rain drove us to seek shelter in some huts at a short distance, where, though we had comfortable covering, we were forced to keep a strict watch all night, the situation being very much exposed, and completely commanded by a woody hill just hanging over it. On the morning of the 18th, we again set out at daylight, halted to breakfast at Panawortahcottah, an old fort, four miles distant. . . . Proceeding again after breakfast, we marched through a country entirely impassable in the rains, consisting of dreary swamps and steep hills, but not so much jungle as to the northward, and reached a post called Poorakaudy, twelve miles farther, where, intending to halt for a few hours, we had just ordered tiffin, and the men were beginning to boil their rice, when word was brought us that the rebels, in force, were in the act of besieging another post. Leaving a small guard with our servants and supplies, we pushed on again at such a rate that in two hours and a half we got over ten miles, and reached the out-guards of the rebels, who fled immediately. We pursued as hard as we could run, in hopes of being in time to come on their main body, but they were on the move when we reached the neighbourhood of the post of Gunnypuddy Wuttum, or Sooltaun Battery, having been erected by Tippoo Sultan to awe the people of that part of the country. A running fight ensued, and we soon perceived that they had the best of it in agility. On our return to the battery, we found they had actually commenced filling up the ditch with bundles of straw, whilst the garrison (30 men of the 2nd Battalion, 3rd Regiment, under a native officer), having expended all their ammunition, were silent spectators of the progress of a work which must, in a few hours, deprive them of their lives. Their joy on seeing us may therefore be easier conceived than expressed. The enemy, fully aware of all our previous movements, had traced us to Panawortahcottah in the morning, and concluded that it would take us at least two days to reach this place—all former forces acting in this intricate country having moved with camp equipage and baggage at a rate of from seven to eight miles a day, whereas we had not even a horse amongst us, but all walked, carried our arms, etc., and slept on the ground. We found this a very well-built redoubt, on a slight eminence, commanding a small but neat village and bazaar, with the main road passing right through it, and well stored with grain and provisions. We had therefore only to furnish the garrison with ammunition to restore matters in this quarter. We took a long ramble the next day in search of the rebels, but all in vain, since they had dispersed, never to assemble in arms again, and we returned the following morning to Poorakaudy. Here we enjoyed some rest and refreshment; and I learned that this had been an officer's post, with a company, in former days. The rebels had nearly destroyed it for iron, as they did the bridge at Sangaloo, and every other bridge in the country, making arrow and spear heads from their spoil. . . . On the 20th April, our party being refreshed, we set out on a hunting excursion, in hopes of falling in with some of the fugitives. We went first to Eertee Compah, then to Cheengary, then round by Mootil, destroying the habitation of a rebel chief, and returned to Poorakaudy by two o'clock p.m., without success. . . . All the Coormers, a particular caste of the natives, some of whom had, through terror, joined the Coorchers, or rebel archers, having come over to us, they paid their respects and returned to their own farms. Captain James and a party under him having joined us this evening, the next morning we left them at three o'clock, and proceeded through some thick jungle till daylight, when, coming into a high road, we reached Panawortahcottah, thirteen miles distant, about 8 a.m. There we breakfasted, and then resumed our march by a circuitous route to a pagoda on the bank of the river, in hopes of seeing some of our noble friends the Coorchers. . . . After resting for a few minutes here, we resumed our march, and arrived at Manantoddy at 2 p.m., having walked thirty miles since morning, through unfrequented paths and deep jungle."

On the 22nd April a chain of posts was established to the south-eastward, one at Panawortahcottah, under Captain James, with Lieutenant Rehe and 100 men; the second at Poorakaudy, under Captain Stevenson, with Lieutenant Rule and 50 men; the third under Lieutenant Swayne, at Paukum, with 50 men; the fourth at Moodramole, under Lieutenant Dawson, with 50 men; the fifth at Gunnypuddy Wuttum, with 30 men, under a native officer. All these detachments were furnished by the 3rd Regiment and 1st Battalion, 13th Regiment, the whole being under the general command of Captain James.

On the 23rd April, all the European troops returned to their respective stations.

On the 25th April, Major Welsh marched with 100 sepoys, Captains Pepper and Stewart, and Lieutenants Williamson, Fife, and Meredith, at half-past three, reaching Mooderary, a small village, at 5 p.m., where, hearing no news of the rebels, they halted for the night. On the 26th, this detachment moved to a post called Wallaat, six miles distant, and at ten o'clock marched on in a westerly direction, and halted at a good range of houses on the brow of a hill. Whilst sitting at dinner, a party of Colkars came to Major Welsh with the head of Pooricawittle Canialary Cunnien, one of the principal rebel chiefs, whom they had killed in the jungle to the southward, but all his companions had escaped.

At half-past eight on the morning of the 27th, the detachment started, and, after passing over several woody hills and through a very deep defile of nearly two miles in length, arrived at the Moplah village of Coniote, situated in an open spot near the range of Ghauts leading to Tellicherry. Here intelligence was gained of the body of insurgents, of whom they were in pursuit, having attacked a jemadar's party of the 5th Regiment that day in the Cotiaddy Pass, six miles off. Captain Pepper and Lieutenant Meredith, with forty-five men, at once proceeded to the top of the Ellacherrum Pass, about two miles to the west, whilst the remainder, marching by the old post of Martelot, now destroyed, to the top of the Cotiaddy, encamped for the night, after having posted a small guard in the road, and another half-way between, to keep up the communication. At ten o'clock p.m. two shots were fired at the sentries, which were immediately returned, whilst the remainder of the men descended the hill as rapidly as they could. All was quiet again in a few minutes, and everyone enveloped in a thick, dense fog, which wetted everything through, and did not disperse until eight o'clock the next morning. Captain Pepper, who had been joined on the previous night by two officers and eighty men from Manantoddy, was to have gone down the Ellacherrum Pass on the morning of the 28th, and, circling round, to have come up the Cotiaddy Pass, driving the rebels into our teeth. But the rebels were too alert, for, having watched the movements of both parties on the previous night, they had gone off in another direction before daylight. Major Welsh, being uneasy for the safety of Captain Pepper and his party, returned to the post of Martelot, three or four miles distant, and at 6 p.m. was joined by Captain Pepper's detachment, which had made a most fatiguing march of nine hours, over hills on which they could not even trace a footpath, and through the deepest jungle, without seeing anything of the enemy. It was then presumed that the insurgents had fled to the Periah Pass, more to the northward. On the 29th, news arrived that Captain James had captured two rebel chiefs near his post, and Major Welsh returned to Manantoddy, sixteen miles off, which he reached at sunset.

On the 1st May, the head of Ramanumby was brought in to Manantoddy.

Matters being so far adjusted, the forces commenced on the 2nd May to return to their respective cantonments. A great number of officers and men were prostrated by fever, but, with the exception of those already mentioned, there were no casualties. The 2nd Battalion, 3rd Regiment, arrived at Cannanore about the 8th May.

In 1812, Major Welsh, the Commandant of the 1st Battalion of the Regiment, was allowed to train it as a Light Infantry Battalion. "It was rather hazardous at that time to introduce any new system, in deviation from the established 'slow time' of General Dundas, the founder of English discipline; but Sir Samuel Achmuty being just returned from actual and active service, he consented to review the corps, and immediately established four Light Infantry Battalions. I mention this circumstance, because, as His Excellency graciously told me on the parade, it was the origin of light corps in the Madras Army." Major Welsh accordingly received orders to form a light corps from both battalions of the regiment, with permission to select both men and officers from the 2nd Battalion at Cannanore. He arrived at that station on the 27th October 1812, and was "most hospitably received by his friend Colonel Webber, whose corps he came to pick. He soon completed his drafts of officers and men, taking of the latter only unexceptionable volunteers, of whom he found abundance willing." About the 15th December, the drafts for the 1st Battalion, consisting of Lieutenant Fife and five other officers, and one hundred and twenty men, left Cannanore *viâ* the Wynaad, to join their "Light Bob" brethren in Bangalore. A draft of equal strength to replace them, had arrived in Cannanore at the end of October.

1813 On the 25th April 1813, the battalion left Cannanore for Vellore, arriving at that station on the 3rd June. Small guards were furnished by the battalion for the hill forts in the neighbourhood.

1814 From Vellore the battalion marched to Madras, which was reached on the 4th June 1814.

That nine British officers of the 3rd Regiment died between the dates of 18th February 1812 and 25th April 1814 proves how unhealthy the district was in which the regiment had been serving.

1815 Intelligence of the conclusion of the Treaty of Paris having been received in Madras in January 1815, royal salutes with three volleys of musketry were ordered to be fired at all the principal stations of the army, in honour of the event, and batta to be issued to European and native troops.

In 1812 the British possessions in India were first invaded by a large band of Pindarees. These men formed large bodies of irregular horse, and were chiefly distinguished from other troops of the same description, by serving without pay, on condition of being permitted to compensate themselves by plunder. The whole body were mounted, some so well as to form an efficient cavalry, but the far greater part very indifferently on small horses or ponies, and with arms of a miscellaneous description, including pikes, clubs, and sticks pointed with iron. Carrying no baggage, because they trusted to the expedition itself for the supply of their wants, they moved with great celerity towards some previously appointed rendezvous, from which, as a centre, they spread over the whole country, and made a thorough sweep of everything which was portable and possessed any value. As they were not disposed to risk an encounter with regular troops, they endeavoured to fall by surprise on each district marked out for plunder, and to complete the work of devastation before there was any danger of being overtaken. In carrying out this plan, no time could be lost, and hence, as the speediest means of extortion, every species of torture and abomination was resorted to. To check the incursions of these hordes, which had now become frequent, a line of posts from Bundelcund to Cambay was established, but the Pindarees repeatedly broke through, and carried on their ravages simultaneously in all three Presidencies.

A large band carried their depredations to the south and east, entered the Northern Circars, and carried off a rich booty from the district of Masulipatam. In the early part of 1816, the devastating hordes mustered in the greatest numbers they had yet displayed. One of their divisions of five thousand men burst into the territories of the Nizam, and continuing onward, penetrated to Guntoor and Masulipatam, and for eight days kept moving about at the rate of thirty or forty miles a day, committing fearful devastation, and perpetrating horrible atrocities. On the 3rd March 1816, the 2nd Battalion, 3rd Regiment, left Madras for service in the Northern Circars, where it was employed in pursuit of the various bands of Pindarees. After the retirement of the Pindarees, the battalion garrisoned Amaravatty from the 24th April 1816 to the 18th August 1816, and Ongole from the 28th August to the 16th November. On the 22nd November 1816, the

1816

battalion arrived at Datchapillay, where it remained 1817 until the 10th March 1817.

The comparative impunity with which the Pindarees had escaped in March 1816, tempted them to return in December. The population, despairing of being able to offer any resistance, fled to the neighbouring hills and thickets, and left their villages and homes at the mercy of the marauders, who had partially plundered the town of Ganjam, when the approach of the troops hastened their departure. They were not allowed, however, to escape so easily as before. "One British detachment hanging on their rear, repeatedly came so near as to inflict severe punishment on the main body; other detachments intercepted them in their retreat; and when at last they reached their cantonments, it was with greatly reduced numbers, and the loss of much of their ill-gotten booty."

The Pindarees, though the most numerous and most atrocious, were by no means the only depredators. Depredation in some form entered largely into the military system of the Mahrattas, and many of the troops, professedly belonging to Scindia and Holkar, were marauding mercenaries, who trusted much more to plunder than to regular pay, and were ever ready, when dissatisfied with the one or other, to change masters, or to assume independence and create disturbances, merely for the purpose of profiting by them. In order to extirpate this predatory system, the armies of Bengal, Madras, and Bombay took the field in 1817. The British Army amounted to 116,464 men and 295 guns (including contingents), and was the largest ever assembled in India by the British.

The 2nd Battalion, 3rd Regiment, was not attached to any of the brigades destined for offensive movements against the enemy, but was employed from the 10th March 1817 to the 17th October 1817 in protecting Humpsaghur and Masulipatam from their incursions.

From Masulipatam on the 18th October 1817, the battalion joined the force under Major Foulis of the 1st Cavalry, and proceeded to Guntoor and the Palnad, to secure these districts against the depredations of the Pindarees. The force, consisting of whole or parts of eight regiments, was engaged in this service until the end of March 1818, when, in consequence of the dispersion of the several Pindaree hordes, the force was broken up, and the 2nd Battalion, 3rd Regiment, marched for Bellary, arriving there on the 25th July 1818.

1818

On the 9th February 1819 the battalion left Bellary, marching *viâ* Badamy and Huttony for Bijapur, where Brigadier General Pritzler, then commanding the Field Division of the Madras troops in the southern Mahratta country, was engaged in reducing the fort of Copaul Droog, which was held by a rebel vassal of the Nizam. Before, however, reaching Bijapur, the fort had been taken, and the battalion, arriving on the 25th May 1819, went into garrison there until the 30th November 1819.

1819

The battalion afterwards marched from Bijapur, and occupied Kaladghi on the 11th February 1820. Captain Alexander Stewart died at this station on the 10th June 1820, and his death was recorded as follows in the *Asiatic Journal and Monthly Register* :—

1820

"In camp at Kallodghee, after a severe illness, which he bore with much fortitude, Alexander Stewart, Esq. of Stenton, late Captain Commanding the 2nd Battalion, 3rd Regiment. This lamented officer received a severe wound in action with the enemy in 1811, from the effects of which he continued to suffer throughout the remainder of his life. He was actively employed during the whole of the arduous campaigns of 1803–4 in the Mahratta country, under Sir Arthur Wellesley (the present Duke of Wellington) and the late Colonel Wallace. Strict and honourable in his principles, his conduct was marked by an anxious and ardent zeal to discharge efficiently every duty entrusted to him."

During 1821, cholera broke out at Kaladghi, and

1821

the 2nd Regiment Light Cavalry suffered very considerably from it. "It is rather extraordinary, but worth remarking, that the 2nd Battalion, 3rd Regiment, and 2nd Battalion, 19th Regiment, and artillery, which are encamped within less than half a mile of the cavalry, but are more sheltered from the westward, have escaped this dreadful disorder."

The battalion marched from Kaladghi on the 18th October 1821, and occupied Sattarah from the 4th November 1821 to the 18th April 1822, and again Kaladghi, from May 1822 to February 1823, after which it marched for Bellary, arriving there on the 1st March 1823. Leaving Bellary on the 30th March 1824, the battalion went into garrison at Gooty on the 3rd April 1824, furnishing a detachment at Cuddapah under the command of Lieutenant John Sheil.

1822

1823

1824

It was at Gooty, on the 6th June 1824, that, orders having been received to do away with the double battalion system, which had been in vogue for nearly twenty-eight years, the 2nd Battalion, 3rd Regiment, once more became the 13th Regiment of Madras Native Infantry. No reason for this change was assigned by the Court of Directors, but "the effect of the separation was to make promotion run in a single regiment or battalion, instead of in two, otherwise there was no material alteration, inasmuch as the old battalions, practically, and to all intents and purposes, were nothing else than separate regiments having no connection with each other beyond the number." The officers were posted alternately between the 3rd and 13th Madras Native Infantry, that is, all the odd numbers of each rank to the first, and the even numbers to the second battalions of their present regiments. The 13th Regiment Madras Native Infantry was thus numbered in the order in which it was first raised.

Major - General and Colonel W. Macleod was

appointed "Colonel," and Lieutenant-Colonel F. P. Stewart "Lieutenant-Colonel Commandant," of the regiment. Lieutenant G. Dods was appointed Adjutant, and Lieutenant J. Briggs, Quartermaster, Interpreter, and Paymaster.

The following is a copy of the General Order, dated 1st June 1824, notifying the above change :—

"The Honourable the Governor in Council directs that the Regiments of Infantry on this Establishment shall hereafter be numbered as follows :—

Present Number.	*To be Numbered.*
. . . . . .	. . . . . .
2nd Batt., 3rd Regiment, Native Infantry.	13th Regiment Native Infantry.
. . . . . .	. . . . . .

"The above corps, notwithstanding the alteration in the numbering and Designation of Regiments, are each to preserve such Honorary Badges and Devices in their respective colours and appointments as may have belonged to each under its former number as a Battalion of a Regiment.

"Officers are posted to regiments as follows :—

. . . .

"13th Regiment Native Infantry,
Late 2nd Battalion, 3rd Regiment, Native Infantry.

Major	G. Hunter.	Lieutenant	C. Fladgate.
Captain	J. Wilson.	"	J. C. Glover.
"	A. H. Colberg.	"	J. Sheil.
"	J. Fyfe.	"	J. S. Sherman.
"	R. Inverarity.	"	T. G. E. G. Kenny.
"	J. G. Rorison.	"	J. F. Leslie.
Lieutenant	G. Dods.	Ensign	G. W. Watson.
"	J. Robins.	"	J. Everest.
"	J. Briggs.	"	H. C. Beevor."
"	E. Rogers.		

On the 8th August 1825, The Rules and Regulations for the Field Exercise and Evolutions of H.M.'s Forces, dated Horse Guards, 10th March 1824, were adopted by the regiment. 1825

From Gooty the 13th Madras Native Infantry marched, on the 16th December 1826, to Secunderabad, arriving there on 7th January 1827, and, after remaining until the 9th June 1829, garrisoned in turn Cuddapah, from the 17th July 1829 to 7th May 1832; Vellore, from the 25th May 1826 1827 1829 1832

1835 1832 to 20th June 1835; and Madras, from 29th June 1835 until 6th October 1835. From the latter station the regiment first proceeded on foreign service, the headquarters embarking on the 7th October 1835, and arriving at Maulmain, in Burma, on the 2nd November, the remainder of the regiment, under the command of Lieutenant W. F. Goodwyn, embarking on the 26th November and arriving about the end of December. The regiment, on arrival, furnished a detachment at Amherst under the command of Lieutenant J. W. G. Kenny.

1837 In 1837, civil war broke out at Ava between two of the principal nobles of the country. For a long time past the king had been incompetent to manage the affairs, and the queen and her brother, named Tharrawaddy, directed the government. The latter determined to appeal to arms. The following is an extract from a letter dated Maulmain, 1st May 1837, published in the *Asiatic Journal and Monthly Register* for November 1837, concerning the despatch of a detachment of the regiment to Ava, etc. :—

"A detachment from Prince Tharrawaddy's army has taken possession of Rangoon. Our communication with Ava is cut off, and we have no news of Colonel Burney (the British Resident at Ava) since 4th March, and no news of a detachment of the 13th Regiment, a subaltern and thirty men, which left Rangoon for Ava in the beginning of March. On the 26th, the Commissioner here sent Lieutenant Smith, 13th, five sepoys and a dozen of Lascars, to Rangoon, with directions to push on to Ava, with a despatch for Colonel Burney. We are very uneasy from this want of intelligence. The native regiment here is in a state of great discontent, to use a mild term, at the clipping legislation of the Government. Government pledged themselves that it should enjoy the same advantages as the corps it relieved. To-day they have reduced a water-carrier to each company, leaving but one to a company, which is a breach of contract stoutly resented. If a government can gather anything from experience, the Indian one may guess what a dangerous affair it is to coquet with a mercenary's pay. The habit of breaking faith has become second nature to it; and I daresay 'they know not what they do.'"

Lieutenant Horatio Clarke Beevor, who commanded the escort of the thirty men referred to in the preceding paragraph, joined the British Resident at Ava on the

13th April, and returned with the escort to regimental headquarters at Maulmain on the 17th July.

On the return of Lieutenant Josiah Smith to Maulmain, the Brigadier-General commanding published an order "that he was happy to be enabled, by a note which he had received from the Commissioner of the Provinces, to place on record that Lieutenant Smith, 13th Native Infantry, in the discharge of the duty lately entrusted to him of effecting a communication with the Resident at Ava, had performed an important and difficult service with great zeal, promptitude, and intelligence" (*District Order, dated Maulmain, 17th June* 1837). The Governor-General considered the service performed to be highly creditable to Lieutenant Smith, of having effected, at an important juncture, a communication with the Resident at Ava, as brought to notice in *India Political Dept. Letter No.* 10, *dated* 11*th July* 1837 (*India Office Records*). This officer, in recognition of his services, was shortly afterwards appointed to the command of the newly-raised Talain Levy.

The discontent, owing to the reduction of a water-carrier per company, noted in the concluding portion of the extract from the *Asiatic Journal and Monthly Register*, resulted in the mutiny of the grenadier company commanded by Lieutenant George Cumberland Hughes, but was speedily quelled by that officer, whose conduct was highly commended by the authorities (*India Office Records*).

During the disturbances in Burma this year, no loss was sustained by the British, whose property had been protected by the influence of Colonel Burney. A new government for Ava was organised under Prince Tharrawaddy.

The tour of foreign service of the regiment lasted until 28th April 1839, when it re-embarked for India, being relieved at Maulmain by the 31st Light Infantry.

1839

Having arrived at Pallaveram on the 25th May 1839, the regiment remained in garrison there until the 17th August of the same year, when orders were received to march to Vellore, furnishing on the march the relief of the detachments stationed at Arnee and Congeveram, in relief of the 48th Native Infantry. The regiment arrived at Vellore on the 23rd August 1839.

Vellore was garrisoned by the regiment until the 19th December 1840,[1] when it marched to Samulcottah, arriving there on 9th February of the following year. A detachment of the regiment under the command of Captain J. C. Glover was quartered at Masulipatam.

1840

The regiment remained in garrison at Samulcottah until 12th October 1843, after which it was quartered in Chicacole from the 1st November 1843 until 20th March 1845, in Secunderabad from 8th May 1845 to 15th January 1849, and in Cuddapah from the 10th February 1849 until 1st September 1851.

1843 1845 1849 1851

The 13th marched to Trichinopoly from Cuddapah, which was reached on the 6th October 1851. A detachment for Tanjore was furnished, which rejoined regimental headquarters in October 1852 on being relieved by the 29th Native Infantry. The headquarters and right wing of the regiment left Trichinopoly on the 3rd October 1854, reaching Palamcottah on the 23rd. The left wing left Trichinopoly on the 28th January of the following year, and rejoined headquarters on the 17th February.

1852 1854

1856 The turn of the regiment for foreign service having again come round, the headquarters and seven companies embarked at Tuticorin, on board H.M.S. *Syddy*, on

[1] As requisition had been made to the Madras Government for two regiments of native infantry for service in Sinde, it was stated that the 13th Native Infantry would be ordered immediately to that place, "at the prospect whereof every individual in the regiment was much excited," but the Madras Government finally replied that there was no available force in the Presidency for that purpose (*Asiatic Journal*).

the 17th December 1856, and reached Maulmain for a second tour on the 16th February 1857, the other three companies arriving on the 14th March. Detachments were furnished to Sittang, Mergui, Penoung, and Thanoi.

1857

Whilst stationed in Maulmain, the regiment volunteered to serve against the mutineers in the Bengal Presidency, which was acknowledged in the following Gazette, dated Fort St. George, 4th September 1875:—

"No. 277 of 1857.—The Governor in Council directs it to be notified to the army, that the under-mentioned corps have volunteered their services for employment in Bengal.

. . . .

"13TH REGIMENT NATIVE INFANTRY.

. . . .

"His Lordship in Council desires that the thanks of Government may be conveyed to the whole of the native officers, non-commissioned officers, and men."

On the arrival of the regiment in Maulmain, a detachment of three companies, with Captain Brooking, Lieutenant Boardman, and Lieutenant Strettel, was sent to Sittang to relieve a detachment of similar strength furnished by the 8th Madras Native Infantry. This detachment furnished a sub-detachment of 30 rifles, under the command of Subadar Govindoo, to Ky-ka-ti Post, on the Sittang River. About this time Moung Thelah, the late Myook of Ky-ka-ti, who had been imprisoned in the jail at Shwegyn for having made away with Government money, having effected his escape from jail, had collected about 200 Burmans, armed with muskets, lances, and dahs, and had commenced plundering the neighbouring villages. Early in March 1858, news having been received that this Dacoit leader, Moung Thelah, and his men, were going to attack the Ky-ka-ti Post, Subadar Govindoo, the Commander of the Post, after taking such steps as he considered necessary for their reception, applied to Captain Brooking at Sittang for reinforcements.

At about 3 a.m. on the 4th March, the Dacoits, led by Moung Thelah himself, attacked the post. Fighting continued for three hours, after which the Dacoits retreated, and took refuge in a pagoda. The Dacoits were pursued by a portion of the detachment from the stockade; but, finding that an advance over the open country to attack their strong position would be very costly, Subadar Govindoo withdrew to another position in a pagoda immediately opposite the pagoda held by the Dacoits. Fire was opened on the Dacoits, but the ammunition of the detachment running short, the Subadar called for volunteers to bring a box of ammunition which had been left behind during their retirement, and which was lying on the ground between the two pagodas. Havildar Mootooveeru, volunteering for this duty, gallantly ran out and brought in the ammunition, during which time he was exposed to heavy fire from the Dacoits.

The ammunition of the detachment being thus replenished, fire was again directed on them. After a stubborn resistance, the Dacoits fled from the pagoda into the jungle. Here, however, they fell in with the night guard, composed of seven privates, under the command of Naick Venket Doss, which was furnished for the protection of the markets at Khyto, and was returning to the stockade.

The Naick immediately attacked the Dacoits, pouring volleys into them. After making a slight stand, the Dacoits fled. Moung Thelah and five other Dacoits were left killed on the ground, others being carried off by their comrades. Three men of Naick Venket Doss's party were wounded, viz. Private Vithilingam, Private Shaik Mohideen, and Shaik Homed. A brass bullet was extracted from Private Vithilingam's side, but he continued serving until 1869, when he was pensioned after twenty-five years' service. Private Shaik Mohideen's upper arm being fractured by a bullet near the left shoulder, he was pensioned on Rs.7 a month. Private Shaik Homed, who was slightly wounded in the sole of the right foot by a bullet, was promoted Lance-Naick. Reinforcements, under the command of Lieutenant Boardman, arrived at the stockade in the morning after the fight was over. Moung Thelah's head was severed from his body, and despatched to Sittang, where it was hung up for a day, and then buried. The wounded Dacoits were taken prisoners and treated in the hospital.

For this service Subadar Govindoo received the 2nd Class Order of British India as supernumerary to the establishment. The Government presented the men of the detachment with Rs.1000, which was distributed according to the rules regulating the distribution of prize money.

The following are some of the names of the privates who were engaged at Ky-ka-ti, near Sittang, on the 4th March 1858, against the 200 rebels, when the rebel Boh Moung Thelah was slain:—No. 92, Subbiah; No. 1311, Ramasawmy; No. 1354, Shaik Tippoo; No. 1378, Moideen Cawn; No. 1590, Curpremah; No. 1615, Shaik Dawood; No. 1623, Shaik Furreed; No. 1658, Jewan; No. 1667, Combaconum; and No. 1708, Mohommed Sittar.

Lieutenant Strettel, with a small detachment of the 13th Madras Native Infantry, had accompanied the Commissioner of Sittang on "dour," on the 10th December 1857, with the object of capturing Moung Thelah, but without success.

The following is a copy of the Gazette of Fort St. George, 19th October 1858:—

"No. 406 of 1858.—The following General Orders by the Honourable the President of the Council of India in Council, are republished for the information of the army:—

"GENERAL ORDERS BY THE HONOURABLE THE PRESIDENT OF THE COUNCIL OF INDIA IN COUNCIL.

"FORT WILLIAM, *28th September* 1858.

"No. 1355 of 1858.—The Honourable the President of the Council of India in Council is pleased, as a special case, to admit Subadar GOVINDOO of the 13th Regiment Madras Native Infantry, to the 1st Class of the Order of British India, with the title of Sirdar Bahadoor, from the 8th August 1858, in consideration of his conspicuous gallantry in the defence of his post at Ky-ka-ti, in Martaban, against a large body of rebels, headed by the chief 'Moung Thelah.'

"This native officer is to be borne on the list as a supernumerary until a vacancy occurs."

The following is a copy of Adjutant-General's Letter No. 2719, dated 1st September 1858 :—

"TO THE OFFICER COMMANDING THE PEGU DIVISION.

"SIR,—I have the honour, by order of the Commander-in-Chief, to acknowledge the receipt of your letter, dated 29th ultimo, No. 178, transmitting copy of one from the Secretary to Government of India, dated 11th idem, No. 2043, conveying sanction for disbursement of the reward of 1000 rupees for Moung Thelah to Subadar Govindoo of the 13th Regiment Native Infantry, and the party under his command.

"Lieutenant-General Sir P. Grant cordially congratulates the Subadar on the acknowledgment of his good services.

"Adjt.-Gen. Office, (Signed) . . WOOD, *Lieut.-Col.*
Headquarters, Ootacamund. *Adjutant-General of the Army.*"

On the 21st May 1858 the following men of the regiment on detachment duty in Martaban assisted in the pursuit and capture of certain escaped convicts at Mergui, Burma :—

No. 1509, Private Muhammad Hoosman; No. 1499, Private Abdullah Khan; No. 1518, Private Emom Cawn; No. 1585, Private Sied Abdool Cawder; No. 288, Private Jemall Begh.

The first mentioned having particularly distinguished himself on the occasion (*vide*[1] Letter No. 4, Dy. No. 88 of 1858, dated 27th May 1858, from the Deputy Commissioner Mergui) was promoted Naick.

The regiment left Maulmain on the 24th January

[1] Unfortunately no copy of this letter can be found amongst the records of the regiment, nor in the Deputy Commissioner's office at Mergui.

1860, and arrived at Trichinopoly on the 17th February of the same year. 1860

In 1861 a detachment of the regiment, under the command of Captain W. Boardman, was employed at Tanjore in furnishing guards over the treasury and jewels of the Tanjore Raj, there secured under seal; a duty, however, not well fulfilled, as some of these were stolen, and Subadar Shaik Bram lost his commission through not reporting the suspicion attaching to Private Ismail Sherif of the detachment, whose family were reported to have become suddenly rich by sale of gold brought from Tanjore. 1861

The regiment left Trichinopoly in two parties for Cannanore, to take the place of the 18th Native Infantry, which had been disbanded, the left wing leaving the former place on the 27th May 1864, and reaching the latter on the 10th June. The headquarters and right wing did not leave Trichinopoly until the 23rd December 1864, arriving at Cannanore on the 13th January 1865. 1864 1865

On the 24th October 1865, the system of organization then in force in the other Presidencies being introduced in Madras, the following appointments were made in the regiment with effect from the 1st November 1865:—

Lieut.-Col. E. F. Burton (in Europe)	Commandant.
Major C. Pulley, 37th Native Infantry	Officiating Commandant.
Major C. W. Taylor	Senior Wing Commandant.
Captain (Bvt.-Major) J. W. Rutherford, 47 Native Infantry	Junior Wing Commandant.
Captain (Bvt.-Major) W. Boardman	Officg. Junior Wing Commandant.
Lieut. C. M. Moberly	Adjutant.

The old mess was broken up, and the greater portion of the British officers "divorced" from the regiment. All the regimental appointments were henceforth called Staff appointments. All the officers belonged to the Staff Corps, and were interchangeable between the various Madras regiments. Promotion ceased to depend upon regimental vacancies, but was obtained purely by length of service, so that a subaltern became captain after twelve years' service (subsequently changed to eleven), major after twenty, lieutenant-colonel after twenty-six, and colonel after thirty.

The following is a copy of a letter sent to the Commandant after the visit of Lord Napier to Cannanore :—

"BRIGADE OFFICE, MALABAR AND CANARA,
CANNANORE, *20th November* 1866.

"No. 829.

"MEMO.

"The Brigadier-General has much pleasure in announcing to the officers commanding 13th and 40th Regiments Native Infantry, this, His Excellency Lord Napier expressed himself on the occasion of his recent visit as exceedingly pleased with the general appearance of the two regiments, which showed the great care for, and interest in the comfort and well-being of their men, taken by their officers, as evinced by the well-ordered arrangements of the schools, lines, gardens, etc. etc., and the Brigadier-General begs to thank them for their labours which have led to so gratifying a result.

"The substance of this Memo. will be communicated to the officers, European and native, as well as to the men of each corps.

"By order,
(Signed) "G. A. ARBUTHNOT, *Capt.*,
Brig.-Maj. Malabar and Canara."

1870 In November 1870 the regiment embarked at Cannanore and proceeded to Hong Kong in two detachments, the headquarters and right wing arriving there on the 21st December, and the left wing six days afterwards.

Having remained there for one and a half years, the regiment left in two detachments, the first, consisting of the headquarters and right wing, on the 1st May 1872 1872, the second, consisting of the left wing, on the 5th of the same month, arriving at Pallaveram on the 20th and 21st May respectively.

1873 On the 17th February 1873 two companies were sent from Pallaveram to Madras, being followed on the 22nd by another company, and by the headquarters and one company on the 5th March. Two companies were sent on the 16th March, half a company on the

18th, and the remainder joined regimental headquarters on the 10th April 1873.

In July 1875 the Christian Association Fund was established in the regiment for the benefit of the Christian community, Christians of all denominations being eligible to join the Association. Bonuses on account of certain domestic occurrences, transfer, discharge, etc., were authorised.

On the 17th December 1875 the regiment took part in the review which was held in honour of the visit of His Royal Highness the Prince of Wales, and the following is an extract from District Orders by Brigadier-General Raikes, C.B., Commanding Centre District, dated Headquarters, Fort St. George, 18th December 1875:— 1875

"It is with the highest gratification that Brigadier-General Raikes, Commanding the district, communicates to the troops on parade yesterday, that he received the command of Field-Marshal His Royal Highness the Prince of Wales, in person, to convey to all ranks, through commanding officers, that His Royal Highness was pleased to express his entire satisfaction with the parade, and soldierlike appearance of the troops."

Early in 1876 two men of the regiment were sent to Bangalore to be instructed in army signalling (the Morse system), and on their return to headquarters, a regimental class was formed for the first time, and the regular instruction commenced. 1876

During the stay of the regiment in Madras, Private Maduramootoo of the "A" company was presented with a silver medal by the Commander-in-Chief, Sir Neville Chamberlain, K.C.B., G.C.S.I., in commemoration of the assumption by Her Majesty the Queen of the title of Empress of India. In connection with this presentation, the following is an extract from Regimental Orders by Colonel Dawson, Commandant, dated Madras, 3rd January 1877:— 1877

"No. 1.—The Commandant has great pleasure in publishing to the officers, non-commissioned officers, and soldiers of the 13th Regiment

Madras Infantry, the complimentary words used by His Excellency the Commander-in-Chief, Sir Neville Chamberlain, K.C.B., G.C.S.I., in presenting an imperial medal to Private (No. 14) Maduramootoo of the 'A' company. The Commandant selected Private Maduramootoo on account of his long service and unexceptionable character, and these qualifications were specially brought to notice. His Excellency said to Private Maduramootoo: 'Private Maduramootoo, it affords me great pleasure to present you with a medal awarded by order of Her Majesty the Queen. You have been selected as being the most deserving soldier in the 13th Regiment Native Infantry, having served upwards of forty years, and never had your name entered in the defaulter book. To be thus selected is a high honour of which you may feel justly proud. Every soldier in the army is honoured by the distinction conferred upon yourself. The medal is to you a token of high character, subordination to authority, and of every duty honestly performed.'"

On the obverse of the medal was a bust of the Queen, as Empress of India, with the inscription, "Victoria, 1st January 1877"; on the reverse, "Empress of India," in Persian, English, and Hindustani.

1878 On the 25th November 1878 the headquarters and left wing proceeded from Madras by train, reaching Jubbulpur on the 3rd December. The right wing, leaving Madras on the 20th December, joined regimental headquarters on the 28th.

Negotiations between the British and Russian Governments in 1872 led to a mutual agreement as to the direction of the frontier of Afghanistan, which country was accepted as being a neutral and independent zone. The Amir of Afghanistan, Shere Ali, however, began to feel uneasy, in 1873, at the advances of Russia, and, receiving little sympathy from England, developed an unfriendly spirit towards the Indian Government. Shere Ali dismissed the British Native Agent at Cabul in 1877; and Russia having defeated Turkey, the Indian Government felt bound to act, especially as the Czar had despatched a Russian mission to Cabul. A British mission was sent to Cabul, but was stopped at Ali Musjid by the Amir's orders. An army of 35,000 men was now got in readiness, and an ultimatum was sent to the Amir demanding an apology. This being refused, war was declared. A considerable number of troops was withdrawn from Northern India to take part in the expedition, and several regiments from Madras were sent to join the army, and to garrison certain places near the Indian frontier from which the Bengal regiments had been withdrawn.

On the receipt of telegraphic instructions, the 13th Madras Infantry, under the command of Colonel Strickland, proceeded by rail from Jubbulpur in three parties to Meean Meer, the headquarters; and six companies leaving on the 25th September 1879, one company on the 29th, and the remaining company on the following day, and arrived at Meean Meer on the 29th September and 4th and 5th October respectively.

1879

On the 10th October 1879 instructions were received to despatch a wing to Vitakri in Afghanistan, to relieve the detachment of the 30th Madras Infantry stationed there. Accordingly, two companies of the right wing left Meean Meer the same day, being followed by the other two companies the day after. When the wing reached Jampur (two stations from the frontier), orders arrived for the wing to return as far as Dera Ghazi Khan, as the station of Vitakri had been abolished. Agreeably to these instructions, the right wing of the regiment returned to Dera Ghazi Khan, and remained there in garrison.

In the meantime, orders were received for the whole regiment to proceed to Safeed Sung in Afghanistan, and the right wing was instructed to rejoin regimental headquarters, which it did, reaching Meean Meer on the 31st December 1879. However, the order was afterwards countermanded, and the 9th Punjab Infantry, stationed in Meean Meer, was ordered to proceed thither.

The regiment remained in garrison in Meean Meer until the Afghan War was over, and left that station on the 12th October 1880 by train, reaching Jubbulpur on the 16th of the same month.

1880 The following is an extract from G.O.C.C., dated 6th December 1880:—

"His Excellency the Commander-in-Chief publishes with much satisfaction the following extracts from a letter received from the Adjutant-General in India:

"1. I am directed by His Excellency the Commander-in-Chief in India to express to you His Excellency's appreciation of the services rendered by the Madras troops in this Presidency during the recent operations. 2. The conduct of the regiments employed in the field has been all that can be desired, while the services rendered by the sappers have been highly commended by all officers under whom they served. 3. The regiments also employed on garrison duty conducted themselves to His Excellency's satisfaction."

1882 In 1882 the regiment was designated "the 13th Regiment Madras Infantry," the word "native" being omitted from the title.

1883 In 1883 the Welsh tune of "Ap Shenkin" was adopted as the regimental march past, in supercession of "Blue Bonnets over the Border."

1884 The regiment left Jubbulpur on the 16th October 1884, and marched to Bellary to relieve the 27th Madras
1885 Infantry, arriving there on the 13th January 1885.

King Theebaw, the cruel king of Burma, after recklessly squandering his money in every possible extravagance, and after a vain attempt at an intrigue with the French, endeavoured to replenish his exhausted treasury by demanding "back timber duties" from the Bombay and Burma Trading Company. The agents of the Company refused to pay, and shortly afterwards, some of their employees having been fired upon by King Theebaw's troops, the Viceroy of India despatched an ultimatum, demanding that a British Agent should be received in Mandalay. The king of Burma gave evasive and unsatisfactory replies. The Indian Government made preparations for war, and in November 1885, the ultimatum not having been complied with, despatched a force of about 9000 men, under Major-General H. N. D. Prendergast, C.B., across the frontier. Minhla was occupied after a sharp fight, and shortly afterwards the General received the unconditional surrender of the king and his capital. The whole of Burma was annexed to the Indian Empire, the king himself being despatched under escort to Rangoon. The fall of Mandalay was followed by the disbandment of the Burmese army; many of whom, carrying off their arms, attached themselves to the several pretenders to the throne, who had sprung up in consequence of the unsettled state of the country. They were joined by the ex-officials, the criminal classes, and the disaffected generally. Everyone in power with a grudge against another commenced to pay off old scores; whilst the Dacoit leaders, whose opportunity in Burma has always been in troublesome times, did not neglect to take advantage of the confusion to collect their adherents, and roam about the country on their own account in search of plunder. Flying columns were sent out and patrolled the jungles in all directions, and a system of military posts established throughout the country, to maintain order in their immediate vicinity, and to hold in check the numerous organizations which had sprung up in local opposition to British administration. No attempt was made at co-operation by these organizations. All were totally disconnected, the object of each being its own aggrandisement.

1886

Instructions having been received for the 13th Madras Infantry to proceed on foreign service to Upper Burma, to take part in the above-mentioned military operations, the regiment left Bellary on the 7th December 1886, under the command of Colonel W. Anderson, embarking at Madras on board the B.I.S.N. Company *Nevasa*, and reaching Rangoon on the 19th. Two days afterwards the headquarters and right wing of the regiment proceeded by rail to Tounghoo, the left wing being left behind in Rangoon. From Tounghoo the headquarters and right wing proceeded in small boats up the Sittang River, and reached Ningyan (now called Pyinmana) on the 29th December 1886.

At this time the Upper Burma Field Force, under the command of Major-General G. S. White, V.C., C.B., was divided into six brigades and one independent command.

The 13th Madras Infantry, stationed at Pyinmana, formed part of the 3rd Brigade under the command of Brigadier-General W. S. A. Lockhart, C.B. The 3rd Brigade was chiefly occupied in hunting down and dispersing the bands of Dacoits infesting Yemethin and Pyinmana. Night surprises and ambuscades, which usually took the form of a volley followed by flight into very dense jungle, were the only formidable tactics of the Dacoits.

Shortly after the arrival of the regiment in Pyinmana, reports having been received that Buddha Yaza and Boh Ay, with 400 men, intended to pass through the district north of Yemethin, making for the Shan Hills, all posts concerned were warned, the roads patrolled, and a move in three columns was ordered. All were directed to concentrate at Salay. The 1st column was sent from Yemethin on the 11th January with instructions to march to Thayetpingwin, twenty miles south-west of Salay, where, according to another report, Buddha Yaza and Maung Toke and other Bohs were stated to be in a stockaded position with 3000 adherents and 4 jingals. The 2nd column from Taungdwingyi and Captain Elliot's cavalry from Salay were to act in concert in a simultaneous attack on that position. The 3rd column, consisting of 84 "Queen's," 42 Bombay Lancers, 50 of the 13th Madras Infantry, and 6 Mounted Infantry, under the command of Colonel Holt, Royal West Surrey Regiment, left Pyinmana on the 9th January, with the intention of combining with the columns from

Yemethin and Taungdwingyi in a simultaneous attack on the above position at daybreak on the 14th. Owing to there being several villages called Thayetpingwin, and to the miscarriage of a letter, Colonel Holt's column was misled, and the attack was not delivered. This column, however, did good work by marching through a district till now unsearched by troops, moving by Patlay, Baumagyi, Thayetpingwin, Thabyaing, and Chaunggzo-Baumagyi. Thayetpingwin and Thabyaing had been recently deserted, both being stockaded, the approaches to the latter being blocked, and the surrounding ground spiked with bamboos. Lieutenant Atkinson, who, on discovering this, had dismounted the Bombay Lancers, and was leading his horses by a path into the village, received a severe spike wound in the right leg. The defences at Baumagyi and the villages of Thayetpingwin and Thabyaing were burnt. Colonel Holt's column returned to Pyinmana on the 22nd January. The column from Yemethin reached Thayetpingwin on the 15th after a trying march of thirty-two miles, but the enemy, who had been encamped near, escaped owing to the absence of the guide who had been detailed to lead the Taungdwingyi column, he having been prevented from doing so by a hostile Thugyi (headman of village). The cavalry came up with a small part of the enemy with women and sick, liberated a Madrassi prisoner with a Burmese wife, and captured a brass gun. Buddha Yaza was reported to have crossed into Shan territory on the morning of the 15th.

The left wing of the 13th Madras Infantry reached Pyinmana on the 17th January, and soon afterwards the regiment furnished detachments for the occupation of the under-mentioned posts—

Léwé . .	2 native officers and	100 rifles under Major Leader.
Madawbin .	1 native officer and	40 rifles under Subadar Kannayva.
Thayetgin .	1 „	50 „ (afterwards relieved by 27th Punjab Infantry.)

Pyokgôn .	1 native officer and	50 rifles	under Sub.-Major Sayyid Bahavodeen.
Kanhla .	1 ,,	50 ,,	under Subadar Sevadial.
Gwebin .	1 ,,	50 ,,	,, Subadar Pir Ahmad.
Imbetgôn .	1 ,,	40 ,,	,, Jemadar Narrainsawmi.
Gwanglôn .	1 ,,	30 ,,	under Jemadar Sayyid Mahmud.
Tauk-chanzin	1 ,,	25 ,,	under Jemadar Muhammad Cassim.

When the first expeditionary force arrived in Rangoon in 1885, there were no cavalry or mounted infantry attached to it, but shortly afterwards 150 mounted infantry were formed. These were found to be invaluable for the work in hand; and led to their organization on a larger scale. In 1886 eleven new companies were raised, and after the arrival of the 13th Madras Infantry in Pyinmana, the regiment furnished 2 havildars, 2 naicks, and 46 privates for No. 8, or the Pyinmana company. In the same company there were 25 British soldiers (including one sergeant and one corporal) of the 2nd Queen's and 2nd Somersetshire Light Infantry. The company was commanded by Lieutenant A. J. Richardson, 13th Madras Infantry. The men were attached to their own regiments for pay, discipline, and rations, and were struck off all regimental duties, and placed under the orders of the officer commanding the Mounted Infantry company. The object of raising the mounted infantry was to provide a ready means of transporting infantry soldiers with greater rapidity and less fatigue to themselves than they could march in an exhausting climate, or against an enemy capable of dispersing and retreating faster than the foot soldiers could pursue with success. The ponies were about 12 hands 2 inches in height, and 57 inches in girth. Each man used the rifle and bayonet he had brought with him from his regiment, the former being slung over his shoulder. The men obtained a free issue of one pair of cord pantaloons, one drab serge coat, and one waterproof cape. Service in the mounted infantry was very popular, and as escorts to convoys of provisions and mails, and to officers travelling from post to post, it would be difficult to exaggerate their value.

News having been received at Pyinmana that a large band of Dacoits had fortified and were holding a place called Saikpudaung, and that San Pe and other leaders were in the hills east of Gyobin, two columns were ordered to leave Pyinmana, with instructions to march by different ways, and to attack simultaneously from opposite directions. Accordingly the 1st column, consisting of 2 guns, 100 "Queen's," 40 Native Mounted Infantry, and a detachment of the 13th Madras Infantry, strength as per margin, left Pyinmana on the 3rd February 1887, under the command of Brigadier-General W. S. A. Lockhart, C.B. The 2nd column, composed of 150 Beluchis under the command of Lieutenant-Colonel Sartorius, left the same day.

BRITISH OFFICERS	5
Colonel W. Anderson.	
Lieut. J. H. Smith.	
Lieut. A. J. Richardson.	
Lieut. E. W. Holt.	
Surg.-Maj. G. Holt.	
Native Officers	5
Havildars	6
Naicks	7
Drummers	2
Privates	120

On the 12th February the 1st column came across two Dacoits, one of whom was captured, the other one escaping through the dense jungle in the direction of Saikpudaung. An officer, accompanied by an escort, was sent with the captured Dacoit to reconnoitre that position, which was found to be unoccupied.

However, on the 13th February, the columns having communicated with each other about noon, it was decided that the 1st column should attack from the front, and the 2nd column from the rear, as it was expected that the position would be occupied by the time the troops arrived. Accordingly, on the 14th, both columns advanced against the place, and found it held by a large number of Dacoits, the Dacoit who had escaped having no doubt acquainted them with the approach of the column. They were in a strong position, spiked and stockaded, with precipitous hills on either flank, which were held in force. The jungle surrounding the position was very dense, where the range and precision of the rifle were of little avail.

When the 1st column had arrived within one hundred yards of the position, the Dacoits opened fire, doing no damage except wounding one of the mounted infantry ponies. The extended portion of the 1st column returned the fire, compelling the Dacoits to abandon the position and seek safety in flight. They were pursued by the mounted infantry, but could not be overtaken, owing to the dense jungle and hilly nature of the country; they fell, however, into the hands of the 2nd column, which was advancing from the opposite direction. Three were killed and several wounded, the remainder effecting their escape under cover of the numerous small nullahs in the vicinity. In this encounter with the 2nd column, one of the Beluchis was killed. A strong post having been established at Saikpudaung, Lieutenant-Colonel Sartorius, with a detachment from both columns (which included a party of thirty men with two native officers of the 13th Madras Infantry, under Lieutenant J. H. Smith), started the same day in pursuit of the Dacoits who had taken refuge in Sepettaung. The enemy being driven from this place, took to the jungle again. After burning the headman's house, and destroying much ammunition and plunder which could not be brought away, Lieutenant-Colonel Sartorius's detachment returned to Saikpudaung.

Both columns remained here for five days, until the country round about had quieted down, after which, with the exception of one hundred men of the 13th Madras Infantry, under Lieutenant E. W. Holt, who were left to occupy the post, the troops returned to Pyinmana. Two months afterwards, this detachment returned to regimental headquarters, in consequence of sickness amongst the men, caused by the very unhealthy climate, and the post was taken over by the 27th Punjab Infantry.

1887

The detachment of the regiment which occupied Madawbin Post was attacked by Dacoits on the 3rd February 1887, but they were driven off, though not finally disposed of; and, attacking the post on two other occasions during the same month, were again driven off into the jungles.

Léwé Post, under the command of Major Leader, of which a detachment of the regiment formed part of the garrison, was attacked by Dacoits on the 7th February; and on two other occasions during the same month, when the Dacoits were dispersed.

The headquarters of the regiment left Pyinmana on the 4th April 1887, and marched to Yemethin, arriving there on the 12th of the same month. A band of Dacoits having collected near Kanhla, which was held by a detachment from the regiment, under the command of Subadar Sevadial, a portion of the detachment was despatched against them on the 10th April 1887, putting them to flight, capturing one of their number, and recovering certain articles belonging to the 16th Madras Infantry, which the Dacoits had in their possession.

Subadar Kannayva, who was in command of Chinzu Post, held by fifty men of the regiment, having sent in a report to Yemethin that a large band of Dacoits had collected in the neighbourhood of his post, orders were issued for two columns to be sent in that direction. On the 11th May 1887, a detachment from regimental headquarters, under the command of Lieutenant A. H. Allenby, strength as per margin, formed part of the 1st column under Major Leader, which was despatched from Yemethin.

British Officer	1
Native Officers	2
Havildars	4
Naick	1
Drummers	2
Privates	95
Puckallies	2

The 2nd column, composed of several artillerymen and one hundred Beluchis, left Pyinmana, under the command of Lieutenant-Colonel Sartorius.

The 2nd column having reached Kogwé, the place of rendezvous, about five miles from Chinzu, an hour before the 1st column, attacked the village, which the Dacoits were holding, killing their leader, Boh Thine, and making prisoners of several Dacoits in the village. Boh Thine was a man in constant communication with Buddha Yaza and the Laywun.

Both columns having now effected a junction, and having left a detachment of twenty-five men of the 13th Madras Infantry, under Subadar Jeanah, in the village, started in pursuit of those Dacoits who had effected their escape. They came across them resting in the jungle, but the Dacoits, however, managed to get away through the dense scrub, after several volleys had been fired. Ammunition and several guns were found on the ground where the Dacoits had been encountered. After the men had rested for an hour, the columns returned to Kogwé, and shortly afterwards returned to Yemethin, arriving there on the 16th May. During this expedition a detachment, 13th Madras Infantry, under Lieutenant Allenby, 13th Madras Infantry, made a long march of thirty-four miles on one day.

On the 19th July 1887, Brigadier-General W. S. Lockhart, C.B., having been invalided to Europe, was succeeded in the command of his brigade on the 23rd August by Brigadier-General H. Collett, C.B.

In addition to the columns previously mentioned, the regiment furnished detachments for many flying columns, who did their fair share of hard work, though the details are devoid of interest. On the 28th November 1887 a patrol of the mounted infantry of the 13th Madras Infantry, under Havildar Abdur Rahim, was fired on by a gang of fifteen Dacoits near

Uyin, a village five miles to the south-west of Kyundon Sayyoywa. The patrol charged, and the Dacoits bolted. One sepoy, 13th Madras Infantry, named Muhammad Haneef, was severely wounded.

On another occasion, a convoy under the command of Naick Abdul Latiff was attacked by Dacoits near Thayetgôn, but the latter were driven off.

The following is an extract from G.O.G., No. 376, dated 24th June 1887, republishing General Order, No. 434, of the Government of India, dated 16th June 1887:—

"The Governor-General in Council now takes the opportunity of the return to India of a large portion of the Field Force, to signify the cordial recognition of the Government of India of the distinguished manner in which the army has acquitted itself during a series of arduous and protracted operations.

"His Excellency in Council congratulates the army upon the spirit which has been shown by all ranks in confronting the difficulties inseparable from a campaign, carried on over a vast extent of country, amidst dense jungles and forests, and under great stress of climate,—conditions which have called forth high military qualities. The difficulties and hardships of the campaign have not been overcome without heavy loss; and the Governor-General in Council deplores the death of many brave officers and men, in action and from disease, who have fallen in the discharge of their duty to their Queen and country.

"The Viceroy and Governor-General in Council has much gratification in announcing to the army that Her Majesty the Queen, Empress of India, has been graciously pleased to approve the grant of the India medal, similar to that conferred for the Second Burmese War, with a clasp inscribed "Burma 1885–87," in commemoration of the services of the officers, non-commissioned officers and soldiers, British and native, who have taken part in the military operations in Burma. The medal and clasp will be granted to all the troops who served in the Burma Field Force, and also to such of those belonging to the Lower Burma Division and the Eastern Frontier District as were engaged in active operations, between the 14th November 1885 and the 30th April 1887, both days inclusive.

"The Governor-General in Council has also the satisfaction to notify that Her Majesty's Government have sanctioned the distribution of a gratuity to all the troops who crossed the old frontier and came under the command of the General Officer Commanding the Burma Field Force, or who crossed from Manipur territory into Burma."

On the 16th September of the same year, a bronze "India" medal of 1854, and bronze clasp inscribed "Burma 1885-87," was sanctioned to all authorised followers who accompanied the army in the field during the operations in Burma, between the 14th November 1885 and 30th April 1887.

The following is an extract from the despatches of Major-General Sir G. C. White, V.C., K.C.B., Commanding Upper Burma Field Force, dated 11th April 1887 :—

"The name of Captain J. E. Preston, 13th Madras Infantry, was not included in my former despatch. This officer rendered specially good service on several occasions in the summer of 1886, and was severely wounded when in command at Lamaing, a post for the command of which I had selected him on account of his gallantry and ability as an officer. I recommend him for advancement."

The following is a copy of G.O.G., No. 696, dated 22nd November 1887 :—

"With the sanction of the Government of India, and under the provisions of Clause 48, I.A.C. of 1884, the following promotions are made for services during the recent operations in Burma :

. . . .

"First Grade Hospital Assistant Jenarden Sing to be Senior Hospital Assistant."

This hospital assistant who had served so long with the regiment was afterwards made a "Rai Bahadur."

1888 The regiment remained at Yemethin until the 16th February 1888, when it left for Pyawbwé, arriving there two days afterwards.

Whilst the regiment was stationed at Pyawbwé, a detachment of one hundred and seventy rifles, under the command of Captain J. E. Preston (afterwards relieved by Captain G. A. Welman and Lieutenant A. H. Allenby), was sent to Koni in the Shan States, a very unhealthy place, owing to the prevalence of "beri-beri," from which disease many men of the detachment afterwards died.

A detachment of twenty-five rifles was also sent to Pwéhla.

On the 28th July 1888 the India Frontier medals with clasps for "Burma 1885–87," for 19 native officers and 689 non-commissioned officers and privates, were presented at Pyawbwé by Brigadier-General H. Collett, C.B.

An outbreak of cholera occurred in the regiment at Pyawbwé, and though every precaution was taken to combat the epidemic, seventeen men of the regiment succumbed to this disease. Several men were also attacked with "beri-beri," six of whom died.

1889

On the 1st January 1889 the Regimental Charity Fund was abolished, and the Regimental Savings Fund established in its place, with the object of encouraging the men to save money while in the service, so that on leaving it they might have the means of settling down comfortably in their native place, or, in case of their death, making provision for their families. All ranks were allowed to pay into the fund any sums saved from time to time. Money once paid in was not to be returned except on ceasing to belong to the regiment by transfer, pension, discharge, or death. Subscribers were permitted on certain occasions, such as the death of any relative depending on them, to borrow from the fund a sum not exceeding a month's pay at one pie per rupee per mensem interest, which money ("Charity Branch") was put aside and used for charitable purposes connected with the families of subscribers who were left in destitute circumstances, each case being considered on its merits by a committee of officers.

A rifle club was first established in the regiment in 1889, with the object of providing members with free or cheap ammunition, in order to encourage rifle-shooting, afford facilities for practice for the various rifle meetings, and provide prizes for match shooting. All ranks in the regiment became members.

The regiment left Pyawbwé for Meiktila on the 1st April 1889, and furnished a detachment of two companies at Pyinmana, which afterwards proceeded to Toungoo, Lower Burma.

1890 On the 22nd January 1890 the regiment left Meiktila *en route* for India, and embarking at Rangoon on board R.I.M.S. *Dalhousie*, arrived at Pallaveram on the 3rd February 1890, after an absence of just over three years on foreign service. A detachment of one company was furnished from Pallaveram for St. Thomas' Mount.

The regiment did not, however, remain long at Pallaveram, for, leaving that place on the 25th March 1890, it arrived in Bangalore on the following day in relief of the 12th Madras Infantry.

After arrival in Bangalore, the regiment was placed in the 1st Army Corps in the event of mobilisation.

Shortly afterwards, orders were received to carry out the recruiting for the linked battalions of the regiment (see Part II. year 1886), namely, the 20th and 22nd Madras Infantry then on foreign service in Upper Burma. Recruiting parties were detached from the regiment and sent to Madras, Vellore, and other places for this purpose. Many recruits for the linked battalions were also enlisted in Bangalore.

The following is a copy of the General Order, No. 31, by the Government of India, dated 10th January 1890, republished in G.O.G., No. 66, dated 28th January 1890:—

"MEDALS—BURMA.

"The Viceroy and Governor-General in Council has much gratification in announcing to the army that Her Majesty the Queen, Empress of India, has been graciously pleased to approve the grant of the India medal with a new clasp inscribed 'Burma 1887–89,' being extended to all troops engaged in the military operations in Upper Burma, and to those actually engaged on field service in Lower Burma, between 1st May 1887 and 31st March 1889, both dates inclusive.

"A bronze medal and clasp of similar pattern will be issued to all authorised Government followers who accompanied the troops engaged."

In 1890, messes for recruits, and recruit and pension boys, were first established in the regiment. Any soldier was allowed to join the mess, but all soldiers of less than three years' service, and all recruit boys, and such pension boys as the Commandant directed, were ordered to be members. The object of the institution was to give the members as much good and wholesome food as possible at the lowest possible cost per head, the importance of change of diet being borne in mind, and also to prevent them from half starving themselves to support their numerous relations. The working of the messes was regulated by a committee of three native officers, subject to the supervision of a British superintending officer. One of the committee attended daily each time the meals were prepared, and ascertained that the food supplied was of good quality, properly prepared, and of sufficient quantity. Each man was provided with two good meals a day, and in addition, coffee and hoppers in the morning.

In 1890, four Morris' aiming tubes were purchased, and a miniature rifle-range constructed, in order to improve the shooting of the regiment. Condemned Martini-Henry rifles were used for the tubes, and every member of the rifle club was allowed twenty rounds every year.

The regiment having been served out with complete sets of signalling equipment in 1890, the course of instruction in army signalling laid down in Clause 112 of I.A.C. of July 1889 was carried out. A class of ten non-commissioned officers and men was put through a complete course, under a British assistant instructor. Intelligent men with a knowledge of reading and writing English, who were able to read four words a minute with the small flag, were selected. The course of instruction lasted for ninety days (1st August to 31st October), after which a month's brigade practice with British troops was held. The training and all subsequent practice was carried out in English. Sixteen signallers were to be maintained in the regiment, of whom the eight best were to receive a bonus of five rupees annually. Regimental classes were also held for the purpose of training men as signallers, and of ascertaining those who were best fitted to send to the special classes.

Classes for the practical instruction of the native ranks in military reconnaissance were first formed in this regiment in 1890, and regimental classes for non-commissioned officers and men were afterwards carried on all the year round, in accordance with S.G.O., No. 76, of the 7th December 1888. In order to provide native instructors in this subject, one non-commissioned officer and one private were attached to the Queen's Own Sappers and Miners for one year's instruction. In 1897 two plane tables were purchased, and the men were taught their use.

A Soldiers' Industrial Exhibition was held in Bangalore in September 1890, for the exhibition of articles which had been made during the leisure hours of the soldier. The Exhibition was opened by the Resident of Mysore, Sir Oliver St. John, and prizes to the value of 2500 rupees were distributed. Many men of the regiment exhibited articles and obtained prizes.

The regiment took part in the "assault-at-arms" held at Bangalore in October 1890, and obtained several prizes.

On the 9th October 1890, H. E. Lieutenant-General Sir Charles Arbuthnot, the Commander-in-Chief of the Madras Army, held a levée at the Mayo Hall, when the British and native officers were presented to him by the Commandant, Colonel W. L. Ranking.

1891 In January 1891 a camp of instruction for the troops in the Bangalore District, under the command of Brigadier-General Bengough, C.B., was held, in which the regiment took part. The troops were encamped at Agrahar, Nundydroog, and other places in the vicinity, from the 2nd to the 21st January 1891, during a portion of which time service conditions were maintained, as it was thought that, if this system was carried out during a number of entire days and nights, the actualities of service would be more closely approached than by discontinuing action after a fixed number of hours. The following is an extract from the District and Garrison Orders by Brigadier-General H. M. Bengough, C.B., Commanding the Bangalore District, dated Bangalore the 26th January 1891, which were issued on the conclusion of the camp of instruction:—

"The manœuvres have been carried out generally in a workmanlike and soldierlike manner, and all ranks have displayed a good spirit throughout, and a keen interest in the several exercises. . . . The infantry made some good marches in excellent time, and commanders generally recognised and exemplified the value of flank and counter attacks. . . . The conduct of the troops has been exemplary. . . . The manœuvres were brought to a close by a good forced march to barracks, the infantry covering a distance of some twenty-five miles and fighting a brisk action at the end." (Only two sepoys of the 13th Madras Infantry fell out.)

"The 4th Madras Pioneers carried their complete equipment, and 13th Madras Infantry their coats *en banderole*, and showed that the Madras Infantry still possess at least some of the soldierlike qualities of the past."

On the 10th July 1891, clasps for 1887–89 were distributed by Colonel W. L. Ranking, the Commandant at Bangalore, to 6 British officers, 19 native officers, 689 non-commissioned officers and privates, and 32 followers.

On the 13th November 1891, a synopsis of gymnastic instruction was introduced into the native army, by which the gymnastic training of recruits was made compulsory. All recruits were ordered to undergo a three months' course, the duration of each lesson being one and a half hours, morning and evening. All drilled soldiers of under thirty years of age were ordered to attend a month's course every year.

On the 1st September 1891, the Soldiers' Industrial Exhibition was opened by Sir Harry Prendergast,

C.B., V.C., when several men of the regiment again competed and obtained prizes.

In October 1891 an "assault-at-arms" was again held at Bangalore, in which teams from the regiment were successful in carrying off several prizes.

The 13th Madras Infantry was replaced in the 1st Army Corps by a regiment at Secunderabad on the 25th November 1891.

1892 In July 1892, No. 2250, Havildar Ponnusami compiled a *Catechism of Musketry*, which, being approved of by the A.A.G. for Musketry, was translated into Urdu, Tamil, Telugu, and Nagri. The havildar sold 1130 copies to various regiments in India.

On the 24th October 1892, about 200 recruits, who had been enlisted in Bangalore, Madras, Vellore, and other places by recruiting parties from the regiment, for the 20th and 22nd Madras Infantry, left Bangalore to join the 20th Madras Infantry, under Captain Monck Mason at Secunderabad.

The troops in the Bangalore garrison were reviewed by the Viceroy, Lord Lansdowne, on the 23rd November 1892, the regiment taking part in the parade.

Orders were received for the regiment to leave Bangalore on the 15th November 1892, but in consequence of the visit of the Viceroy the move was postponed until the 25th of the same month, when it left by route-march for Cannanore to relieve the 29th Madras Infantry. The regiment marched *viâ* Seringapatam and other places in Mysore, where nearly a century ago it had done such excellent service, and reached Cannanore on the 15th December of the same year. The families of the native ranks left Bangalore by special train on the 13th December, and arrived at Ferok on the following night. They embarked on board twelve pattimars on the 15th, and reached Cannanore safely on the evening of the 16th.

1893 The following is an extract from Regimental Orders published by Colonel W. L. Ranking on leaving the regiment in Cannanore on the 21st March 1893:—

"The conduct of the men leaves nothing to be desired, and the fact that during the recent march not a single man was brought before the Commandant for punishment reflects great credit upon them."

In 1893 a reading-room and recreation club was established in the regiment, for which all officers and non-commissioned officers were eligible as members, a separate compartment being provided for the commissioned ranks. The principal English and vernacular papers were taken in, and many professional books purchased from the fund. The rooms were open daily from 8 a.m. to 8 p.m.

Lieutenant-Colonel G. C. Fenwick, A.A.G. Rangoon District, arrived at Cannanore on the 10th June 1893, having succeeded Colonel Ranking as Commandant.

Major-General C. J. East, C.B., Provisional Commander-in-Chief, inspected the troops in Cannanore on the 20th September 1893; and a short time afterwards, that officer wrote a letter to the Commandant concerning the inspection of the regiment, which contained the following remarks:—

"All that I have seen of your regiment is very satisfactory. You have a fine body of men, and they are very steady on parade."

The Governor of Madras, Lord Wenlock, held a levée at Cannanore on 20th October 1893, when the British and native officers were presented to him.

1894 On the 24th March 1894, Lieutenant-Colonel G. C. Fenwick left the regiment to take up the command of his old regiment, the 1st Madras Pioneers, in which he had spent most of his service. He was succeeded in the command by Captain H. Lawson of the 14th

Madras Infantry, who arrived in Cannanore on the 26th May.

BRITISH OFFICERS .	2
Lieut. A. H. Allenby.	
Lieut. R. P. Jackson.	
Native Officers .	4
Rank and File . . .	150
Hospital Assistant .	1
Ward Orderly . .	1
Bhistis	2
Sweepers	2

A column of the marginally-noted strength was sent on an expedition against the Moplahs in the south-west of the Malabar District on the 4th April 1894, in compliance with the following telegram from the Collector of Malabar to the Officer Commanding Cannanore, received about 8 a.m. the same date: "Please send two companies Native Regiment to Manjeri *viâ* Calicut with least possible delay. Moplah rising. Bring them by boat from Badagara." The column started at 4.30 p.m. on field-service scale with 200 rounds of ball ammunition per rifle (40 in pouches). Greatcoats were carried by the men. Badagara, distant twenty-eight miles, was reached at 6.15 a.m. on the 5th. Halts were made of three hours at Tellicherry, and two and three-quarters at Badagara. Twenty-five boats, supplied by the civil authorities, conveyed the column by back-water from Badagara to Ellatore, distant twenty miles, where the men arrived at 12.15 a.m. on the 6th, and the march was continued at once to Calicut Railway Station, the total distance from Cannanore, namely, fifty-eight miles (thirty-eight by road and twenty by boat), having been covered in thirty-six consecutive hours without a man falling out. Soon after arrival at Calicut the men were entrained in the ordinary mail train. Arriving at Parli at 1.15 p.m., news was received that the Moplah outlaws (31 in number) had been that morning at 8 o'clock shot down by a detachment of the 1st Dorset Regiment from Malapuram, under the command of Captain Ridley. Orders were received at Parli for the detachment to march to Manarghat, distant nineteen miles, which was reached on the 7th April. On the 9th April, at the request of the Collector of Malabar, the column was split up into three detachments, and stationed at Angadipuram (under Lieutenant A. H. Allenby), Manarghat (under Lieutenant R. P. Jackson), and at Manjeri (under Subadar Kannayya). Shortly afterwards detachments were sent to Allanalur and Pandikad. By degrees, as the excitement abated, Manarghat, Allanalur, and Pandikad were abandoned, one European officer withdrawn, and the detachment located at Manjeri and Angadipuram, where they were housed in school buildings, etc., and remained there the greater part of the monsoon. On the 30th July 1894, Lieutenant A. H. Allenby returned to regimental headquarters, Lieutenant G. C. Burn relieving him. On the 22nd September the column returned by route-march to Cannanore, where they arrived on the 29th of the same month.

On account of the dearness of provisions at Angadipuram, a mess was established there, and supplies procured from Calicut. This was attended with most successful results, and although it was not made compulsory for men to join, most of them did so. A daily ration of meat was supplied, and the monthly cost of the messing did not exceed Rs.3, 8a. per head.

In view of the very unwholesome liquor sold in native bazaars, and of the undesirable nature of the surroundings of the shops where such is sold, it was considered by Army Headquarters that the establishment of canteens within the lines, where good and sound liquor could be obtained at reasonable prices, would be in the interests of the native soldiers. Accordingly, a regimental canteen was established in the regiment, on the 1st October 1894, for the sale of English beer, rum, and aërated waters to the men. The beer was procured from

Leishman & Co., Neilgherry Brewery, and the rum from M'Dowall & Co., Madras, and the price at which it was retailed in the canteen was As.2 per pint of beer, and As.3 per dram of rum.

For 1894–95 the regiment occupied the fourth place on the list of all Indian Infantry Regiments, and third of all Madras Infantry Regiments inspected in signalling—twenty-three, out of twenty-four regimental signallers examined, being reported to be efficient signallers. The usual bonus was obtained by the regiment.

Whilst stationed in Cannanore, the custom of holding 1895 annual sports for the regiment on the 4th May was introduced, to commemorate the capture of Seringapatam on the 4th May 1799, in which the regiment took part.

In 1895 the following were registered as the regimental colours at the Army and Navy Stores, London:—

1 inch scarlet.
$\frac{1}{3}$ inch drab.
1 inch green.

Under telegraphic instructions from Madras Command Headquarters, a detachment of the regiment, strength as per margin, under the command of Captain J. A. Loudon, embarked at Cannanore on the 19th September 1895, on board R.I.M.S. *Dalhousie*, and sailed for Rangoon, arriving there on the 1st October. The detachment was transferred to the R.I.M.S. *Sladen* on the 2nd, and, proceeding up the Irrawaddy, reached Pakokku on the 13th October, and disembarked on the 14th. On the 15th the detachment marched with 150 mules and 16 elephants to Kanhla (7 miles); 16th, to Tabya (13 miles); 17th, Pyinchaung (11 miles); 18th, Pauk (15 miles). In this march the Yaw river had to be forded twice. The river was partly in flood, and all the mules had to be unloaded, and the baggage crossed on elephants. On

British Officers	2
Native Officers	4
Hospital Assistant	1
Rank and File	205
Public Followers	7

the 20th the detachment marched to Nedat (7 miles); 21st, Tebyin (15 miles); 22nd, Pounglon (10 miles); 23rd, Yawdwin (10 miles); and 24th, arrived at the stockade at Mindat Sakan (Chin Hills), where they relieved a similar detachment of the 9th Madras Infantry. Although the march was a trying one, owing to the bad state of the roads, and constantly having to cross rivers partly in flood, only two men fell out.

British Officers	6
Native Officers	9
Hospital Assistants	2
Rank and File	532
Public Followers	17

On the 31st October 1895, the regimental head-quarters, under the command of Major H. Lawson, strength as per margin, embarked at Cannanore on the R.I.M.S. *Warren Hastings*, and sailed for Rangoon, arriving there on the 7th November 1895. The regiment travelled from Rangoon to Prome in two troop trains, and embarked at Prome on board the R.I.M.S. *Sladen* and two flats, reaching Thayetmyo on the 9th November, in relief of the 9th Madras Infantry.

1896

"The Chin Hills, properly so called, are under the control of a Political Officer, and are subject to regular, though not strict, administration. On the Pakokku border the nearer tribes are subject to very slight control, but they are expected to keep the peace among themselves, as well as to abstain from raids, and they are required to pay tribute. At the beginning of 1896 the peace of the hills bordering on the Yawdwin and Pawk subdivisions remained undisturbed. The Subdivisional Officer of Yawdwin, who is in subordinate political charge of the greater part of this tract, went on tour in December of the preceding year, and paid the customary visits to a number of villages, mainly for the purpose of collecting tribute."

13th Madras Infantry.
Captain J. A. Loudon.
Jemadar Ranganayakulu.
40 Rifles.

On the 13th January 1896, an escort, strength as per margin, left Mindat Sakan with the Political Officer, Mr. Ross, to collect tribute from the villages south of Mindat

Sakan. The escort marched on the first day to the banks of the Che Choung, and encamped on the banks of the nullah six miles from Mindat Sakan. Only short marches could be made at first, as the transport coolies, being short-handed, had to make double journeys each day. On the 14th the escort marched to Twyim, distant six miles, the first three miles being along the bed of the Che Choung, which had to be forded twenty-four times, the water being in most places waist deep. The village of Twyim and other villages in the vicinity were found deserted, the inhabitants having fled into the jungle. A halt was made on the following day to enable the Chin policemen to bring in the villagers. On the 16th the column marched six miles to Chandjee over a very bad road, which in places was so precipitous that it had to be traversed by the men crawling on their hands and knees. On the 17th the march was continued along a ridge about 8000 feet high to a camping ground twelve miles distant in the jungle, which consisted of oak, holly, and walnut trees. At night the cold was intense, owing to the frost. On the 18th, another march of three miles was made, the weather still being very cold, with ice on the tanks and frost in the shade all day. On the 19th the escort marched down the ridge to Choung, distant ten miles, and continuing the march along the bed of the Choung for ten miles, halted at the junction of the two Choungs on the following day. A halt for one day was made here to allow the villagers to bring in the tribute. On the 22nd the escort traversed a very bad road for ten miles to the banks of the Mone river, reaching Tymin village, 8 miles distant, on the 23rd; Auk Myen (6 miles) on the 24th; Tounghi (3 miles) on the 25th; and Chindwe (20 miles) on the 26th. At the last-mentioned place the escort halted for two days. On the 29th, Bozainsakan (15 miles) was reached—a very trying march, the first ten miles being all uphill. The escort arrived at Loungshee, distant 18 miles, on the 30th January, and, leaving the Political Officer at this place on the 2nd February, arrived at Mindat Sakan on the 5th.

13th Madras Infantry.
Lieut. A. T. Kirkwood.
Subadar Viraraghavulu.
25 Rifles.

On the 18th February 1896, an escort, strength as per margin, proceeded with the Political Officer to Man Choung to collect tribute, but the Political Officer, having received a report from the Loungshee Myook that one hundred Chins had committed a savage raid on the village of Shwelegyn on the 17th, and that they had carried away twenty women and children as captives, returned with the escort to Mindat Sakan on the 21st. On the following day the Political Officer, with the marginally noted escort furnished by the detachment, proceeded to Loungshee to inquire into the raid, which was found to have been committed by Chinboks from Kyingyi and neighbouring villages, and Yindus from Atetson Pyedaw, under the leadership of a man named Tin Nge. "Their only form of government appears to be an incomplete village system; each village possessing a thugyi, whose title is hereditary, and consequently does not necessarily indicate in its holder a man of influence. The chief sportsmen in each group of villages are held as being most competent to settle disputes, and they lead the people. But when quarrels arise no arbiter is necessary, for the parties resort to deadly combat, the insult or injury, whichever may have been the exciting cause, 'being wiped out with blood only.'" On the 2nd March, Lieutenant Kirkwood, with ten rifles, returned to Mindat Sakan, having left Subadar Viraraghavulu with the remainder of the men at Loungshee to protect that village against a possible Chin raid.

13th Madras Infantry.
Lieut. A. T. Kirkwood.
Subadar Viraraghavulu.
40 Rifles.

On the 4th March, a detachment, strength as per margin, proceeded to Kachoung, sixteen miles distant from Mindat Sakan, for the protection of that village.

13th Madras Infantry.
Lieutenant A. T. Kirkwood.
Jemadar Sikandar Khan.
40 Rifles.

The Loungshee and Kachoung detachments returned to Mindat Sakan on the 14th March, having been replaced by Military Police from Pakokku.

On the 10th March orders were received for a column, strength as per margin, to proceed with the Political Officer, Mr Ross, to punish the Chins who had taken part in the Shwelegyn raid. Owing, however, to the transport coolies (Chinbon Chins from near Loungshee) not arriving until the 18th, the column could not start before the 19th.

13th Madras Infantry.
Captain J. A. Loudon.
Subadar Viraraghavulu.
Jemadar Sikandar Khan.
100 Rifles.

Military Police.
One British officer.
One Jemadar.
50 Rifles.

A reserve of 50 Military Police, under a British officer, was left at Mindat Sakan, with orders to join the column in the event of it having to move into the Chinmé country.

On the 19th the column reached the banks of the Che Choung, arriving at the village of Twyim on the following day. This and the surrounding villages were found deserted. About an hour after the column arrived in camp, a Chin policeman, who had been sent on ahead to bring in the Dap Thugyi, came into camp and reported that he had heard that the Chins had attacked the Mindat Sakan Post that morning. About an hour afterwards, a villager from near Mindat Sakan came and gave the same information. As it was thought probable that Lieutenant A. T. Kirkwood, who had been left behind in command of the post with 98 rifles of the 13th Madras Infantry, in addition to the Military Police in reserve, would not be able to get a runner through,

Subadar Viraraghavulu with 50 rifles was sent to Mindat Sakan, with instructions to Lieutenant Kirkwood to submit a report of the attack on the post. This native officer rejoined the column on the 21st March.

The following is the account of the attack on the Mindat Sakan Post:—

"At 5.45 a.m. on the 20th March 1896, seven or eight Chins came near the sentry on the quarter guard, and the sentry, thinking that they were friendly Chins, opened the gate to let them in, when one of them, having thrown an egg as a signal, they discharged a shower of arrows at the guard. When the arrows were discharged, the sentry turned out the guard, and two of the arrows, passing through the mat wall, wounded Private Rahman Khan, one of the guard, in the breast and thumb. The guard fired on the Chins, who, being reinforced by others who had hidden themselves in the jungle, now numbered about 200 men. Whilst the guard was engaged in front, some Chins succeeded in pulling down a portion of the fence at the north-east corner of the stockade and effected an entrance. These Chins severely wounded Lascar Khadir Khan with a dah (Burmese sword), but the remainder of the garrison promptly turning out, the Chins bolted into the jungle. Lieutenant Kirkwood, hearing shouting in a north-easterly and south-westerly direction, took a party of eight men outside the stockade and fired on the Chins, killing one, and wounding others, who got away. They were pursued for half a mile beyond the stockade. On returning to the stockade, thirteen coolies from the Political Officer's column came rushing in, stating that they had been attacked, and that three of their number were missing. Lieutenant Kirkwood immediately took twenty men and marched out three miles, but he did not fall in with any of the hostile Chins. The probable reason for this outbreak on the part of the Chins was, that they, seeing the strong column set out on the 19th March from the post with the Political Officer to punish the raiders of Shwelegyn, thought that they would find Mindat Sakan without any garrison. Doubtless they were also elated with their success after the Shwelegyn raid. The Chins who had taken part in the attack were Chinboks from the Che Choung, and they were reported to have returned to their villages."

To return to the column with the Political Officer which was encamped at Twyim.

At 8.30 p.m. on the 21st March, the party noted in the margin left camp with the Political Officer to surprise the village of Kyingyi, the villagers of which were supposed to have taken part in the raid on Shwelegyn, and the abortive attack on Mindat Sakan. The party arrived at the village about midnight, after a very trying march, but found it deserted

13th Madras Infantry.
Captain J. A. Loudon.
Subadar Viraraghavulu.
50 Rifles.

with the exception of one house, which was immediately surrounded. Six men were captured. On returning to the place which had been selected as a camping ground, one of the prisoners, who was in the custody of a Chin policeman, jumped down the khud (precipice) and escaped. Shortly afterwards, another prisoner, who was in the custody of Private Sayyid Hussain, also tried to effect his escape in the same manner, but the private holding on to him in the most plucky way, both rolled down the hill for a hundred yards to the Choung at the bottom. The column arrived in camp about 8.30 a.m. on the 22nd March. One of the prisoners stated that the Kyingyi villagers had taken part in the attack on Mindat Sakan and in the raid on Shwelegyn under the leadership of a Chin named Tin Nge. He also promised that if he and one of his sons, who had been captured the night before, were released, he would bring in two of the captives taken in the Shwelegyn raid. The Political Officer sent him out to bring in the captives.

About 2 a.m. on the 23rd March, No. 1 and 2 picquets were fired on by about ten Chins moving across their front. About an hour afterwards, the Chin who had been released returned to camp with the two promised captives. At 5.30 a.m., Captain Loudon with thirty rifles went along the ridge on the north of the camp to try and find where the villagers of Kyingyi had hidden their grain, and to follow the Chins who had fired on the picquets in the early part of the morning. The party returned to camp about 2 p.m., without finding either the Chins or their grain.

13th Madras Infantry.
Captain J. A. Loudon.
25 Rifles.
Military Police.
25 Rifles.

About 10 p.m., information having been received that the Twyim villagers were going to hold a pwé (Burmese play) in honour of the Chin who had been killed at the attack on Mindat Sakan, the Political Officer, with the escort noted in the margin, left for Twyim, which was reached at dawn. During the night the village had accidentally caught fire, and more than half the houses had been burnt down. The remaining houses having been surrounded, the Chins discharged a shower of arrows, wounding one man of the 13th Madras Infantry, after which they fled into the nullah below the village. The escort fired a few rounds, killing one of the Chins. Four villagers were captured, one of whom stated that ten of the Twyim villagers had taken part in the attack on Mindat Sakan. The escort returned to camp at 11 a.m.

The Chin who had brought in the two captives taken in the raid on Shwelegyn, having been promised that Kyingyi would not be burnt, and that his other son would be released, was sent out again by the Political Officer. He returned to camp on the 25th March with the remainder of the captives.

13th Madras Infantry.
Captain J. A. Loudon.
30 Rifles.
Military Police.
20 Rifles.

On the 26th March the Political Officer, with the escort noted in the margin, returned to Mindat Sakan, taking with him the recovered captives and the prisoners from Twyim. The escort halted at Mindat Sakan on the 27th, and returned to Kyingyi on the 28th. On the 30th March the whole column marched to the Kykou toungyahs (certain lands under periodical cultivation on which temporary mat-houses are erected), which were found deserted. A large quantity of grain was destroyed at this village.

13th Madras Infantry.
Captain J. A. Loudon.
40 Rifles.
Military Police.
One British officer.
30 Rifles.

At 3.30 a.m. on the 31st March, the marginally noted column left camp to surprise Mahlini village. When about a mile from this place, the party was divided into two. Captain Loudon's party, consisting of 35 rifles, surrounded Mahlini, which was found deserted. All their grain was destroyed.

On the 1st April the marginally noted parties left camp to search the jungles for the villagers' camps. The Military Police came on a camp which had been just deserted, and captured several pigs, goats, and fowls. After burning several toungyah huts, both parties returned to camp in the evening.

13th Madras Infantry.
Political Officer and Captain Loudon. 30 Rifles.
Military Police.
One British officer.
30 Rifles.

On the following day the party noted in the margin left camp at 5.30 a.m. for a similar search. An old camp was discovered, and after burning some toungyah huts, the party returned.

13th Madras Infantry.
Captain J. A. Loudon.
15 Rifles.
Military Police.
10 Rifles.

On the 3rd April the column marched to Twyim, and early the following morning marched along the ridge above Twyim in the direction of Min-tin. When the column arrived at the end of the ridge, the villages of Kooim and Chookim were seen in the valley below, with Min-tin about two miles farther on. As the villages were occupied, it was decided to camp on the ridge, which was covered with thick jungle, and to surprise the villages in the morning. The column, however, moved back along the ridge for about a mile, in order to obtain water. On the 5th the party marginally noted left camp at 3.30 a.m. for Kooim and Chookim. Owing to the road being very bad and much longer than was supposed, the party did not arrive until 6.30 a.m., and the villagers, seeing the column moving down to the Choung, began to fly from the villages into the jungles. Some of the Military Police fired on about twelve Chins moving up the opposite hill, killing two and taking four prisoners. Several bullocks, goats, and fowls were captured. One of the prisoners was released in order to try and persuade the villagers to return to their

13th Madras Infantry.
Captain J. A. Loudon.
35 Rifles.
Military Police.
One British officer.
35 Rifles.

homes and give in their submission. The prisoners stated that some of the Kooim, Chookim, and Min-tin villagers had taken part in the attack on Mindat Sakan.

On the 6th April the Political Officer, with an escort of the Military Police, visited Min-tin. They came across several Chins in the toungyahs, who said they would call in the rest of the villagers from the jungles. A halt was made on the following day, to allow the villagers to bring in the balance of tribute, and to pay the fine for taking part in the attack on Mindat Sakan.

On the 8th April the column marched to Pwy, which was found deserted. The village of Tepon was also found deserted, but some grain was destroyed. The villagers of Pwy sent word that they would come in and bring in the Tepon villagers. All the villagers arrived on the following day, bringing with them the tribute and fine.

On the 10th April the column returned to Mindat Sakan. These operations had been most successful, as Rs.1500 had been collected as tribute, and a great deal of useful work had been accomplished, though great difficulty had been experienced in getting good interpreters.

The following is a description of how the various tribes, against whom the detachment operated, were armed :—

> "The Chinbok and Yindu carries a bow from his birth upwards, which is made of bamboo well seasoned by being smoked for five or six years over the fireplace in his hut. The bows are four feet long, and the arrows are of bamboo, eighteen inches in length, and about the thickness of a pencil, being iron tipped for use against big game and in war, and of various shapes, barbed or otherwise. The arrows carry one hundred and fifty yards, and will kill a tiger at eighty yards. The only other weapons carried by the Chin is a dagger about a foot long, used for both killing men and for household purposes, and a spear. The Chinbons differ from the Chinboks and Yindus in that they carry nothing but spears and a leather shield when on the war-path."

Captain J. A. Loudon.
Subadar Sikandar Khan.
Jemadar Azhagarsami.
98 Rifles.

On the 27th May 1896 the detachment of the regiment at Mindat Sakan was reduced to monsoon strength, and the party, strength as per margin, left Mindat Sakan on that day, arriving at Thayetmyo on the 6th June. Lieutenant A. T. Kirkwood was left in command of the post.

"The provisional composition by class companies of the existing material in regiments of Madras Infantry was sanctioned with effect from the 1st September 1896" (see Part II.). Promotions of both native officers and non-commissoned officers were henceforth to be strictly confined to companies of their own class. Orders were received for the 13th Madras Infantry to be formed into three companies of Muhammedans, one company of Tamils, one company of Telingas (under elimination), and three mixed companies.

At the Secunderabad "assault-at-arms" held in September 1896, Private Muhammad Ghaus won the 1st prize in the gymnastic competition for instructors, the 2nd prize being gained by Private Augamautu of the 13th. The regiment was also successful in the gymnastic competition for non-instructors, the 1st prize being awarded to Private Muhammad Ismail.

British officer .	1
Native officers .	2
Hospital assistant	1
Rank and file . .	68
Public followers .	3

On the 14th December 1896, a party, strength as per margin, proceeded to Mindat Sakan under Captain A. H. Allenby to relieve the detachment there.

During 1896, nine non-commissioned officers and men passed in Burmese by the elementary standard, and three by the lower standard, each gaining the Government reward of Rs. 180 for the former, and Rs. 360 for the latter.

1897

British officer .	1
Native officer .	1
Hospital assistant	1
Rank and file .	107
Public followers	2

On the 15th January 1897, the marginally noted party proceeded to Mindat Sakan, under the command of Lieutenant G. C. Burn, journeying as far as Pakokku in the Burma Government steamer *Brigand*, to strengthen the post during the cold weather operations.

Lieutenant A. T. Kirkwood's detachment rejoined regimental headquarters on the 3rd February 1897.

On the 21st June 1897, being the 60th anniversary of the accession of H.M. the Queen Empress, Rs.200 were presented by the Government of India for distribution to the men. This memorable occasion was also signalised by the admission of Subadar-Major Jagannath Singh to the Order of British India (2nd class), with the title of "Bahadur," and by the award of a silver medal with gratuity for meritorious service to Band Havildar R. Paul, and of two medals for long service and good conduct with gratuity to two privates.

The insignia of the Order of British India was shortly afterwards presented to Subadar-Major Jagannath Singh on parade at Raipur, and a photograph taken of the regiment afterwards. In making the presentation, Lieut.-Col. H. Lawson, the Commandant addressed the Subadar-Major as follows:—

"In presenting you with a reward for long and faithful service, which has been conferred upon you by your Sovereign, I feel deeply honoured; and in the fact that the Subadar-Major of the 13th Madras Infantry has been chosen to receive one of the six decorations allotted for the Madras Army reflects the greatest credit on yourself, and is a lasting houour to the regiment to which you belong."

Early in September the regiment volunteered for service against the tribes on the North-West Frontier of India who had taken up arms against the Indian Government, and the application was noted at Madras Command Headquarters.

The Standing Orders and Rules of the various Funds and Institutions of the regiment were first printed in 1897.

During the year 1897, a complete suit of civilian dress, comprising a coat, pair of trousers, turban, and pair of shoes, was presented to every man in the regiment out of the profits of the canteen.

Six Singer's military bicycles (model B), with rifle clips, were purchased largely from the same source, for the use of orderlies, etc., of the regiment.

During 1897 three officers and non-commissioned officers passed in Burmese by the higher standard, eight officers, non-commissioned officers and men, by the lower standard, and seven officers, non-commissioned officers and men, by the elementary standard; the whole obtaining the Government reward of five thousand six hundred and forty rupees.

In consequence of the reduction of the native infantry garrison in Burma, the regimental headquarters, under the command of Lieutenant-Colonel H. Lawson, strength as per margin, embarked on the 8th December 1897 at Thayetmyo on the R.I.M.S. *Porpoise* for Rangoon, arriving there on the 13th December 1897, and immediately transhipped to the R.I.M.S. *Canning* for False Point, arriving there on the 18th December 1897. The men and baggage were then transferred to five flats which were towed up the Mahanadi to Cuttack, arriving there the next evening.

No. of companies	7
British officers	4
Native officers and hospital assistants	13
Rank and file	595
Public followers	20

From Cuttack the regimental headquarters proceeded by route-march to Sambalpur (17 marches) *viâ* Angul, on the 24th December, arriving there on the 13th January 1898.

1898

At Sambalpur the left wing, under the command of Captain A. H. Allenby, strength as per margin, remained to garrison the place in relief of the wing of the 2nd Madras Infantry.

British officers	1
Native officers	6
Rank and file	257
Public followers	5

From Sambalpur the remainder of the regiment, under the command of Lieutenant-Colonel Lawson, proceeded on the 16th Jan. 1898 to Raipur, arriving there the next morning, in relief of the 2nd Madras Infantry.

The company which had been left behind at Mindat Sakan under the command of Lieutenant H. W. Marsden, being relieved by the Burma Military Police, embarked at Rangoon on the R.I.M.S. *Dalhousie*, and rejoined regimental headquarters on the 28th Jan. 1898.

Early in this year two detachments were sent from the regiment to Miransha in the Tochi Valley to be attached to the 20th Madras Infantry, one of the linked battalions of the regiment.

EXTRACT FROM THE INDIA ARMY LIST, DATED 1ST JULY 1898.

13TH REGIMENT OF MADRAS INFANTRY.

Alcock—Formed at Madras in December 1776 as the 13th Battalion of Drafts from the 4th, 7th, and 11th Battalions.
Class Company Regiment—3 companies Muhammedans, 1 company Tamils, 1 company Telingas, 3 companies mixed.
Became 2nd Battalion, 3rd Regiment, 1796. Became 13th Regiment N. I., 1824.
"Carnatic"—"Sholinghur"—"Mysore"—"Seringapatam"—"Burma, 1885–87."
Uniform—Red. Facings—White. Lace—Gold.
Puggri—Khaki, white band, khaki fringe.
Raipur. Arrived 17th January 1898, from Thayetmyo.
Wing at Sambalpur.
(Linked with 20th and 22nd M. I.) Regimental centre—Pallaveram.

First Commission.	Rank, Names, and Corps.	Army Rank.	First appointment to Regiment.	Present appointment in Regiment.	Remarks.
	COMMANDANT.				
12/8/1875	Lawson, Major (Temp. Lieut.-Col.) H. (p.s.)	12/8/1895	26/5/1894	26/5/1894	Lv. out of India, p.a., 1 yr., m.c.o., 10th June 1898.
	WING COMMANDERS.				
11/9/1877	Welman, Major G. A. (p.s.)	11/9/1897	19/6/1882	8/5/1893	2nd in comd., 8th May 1893. Offg. Cantt. Magte. Sialkot.
10/3/1883	Loudon, Capt. J. A. (p.s.)	10/3/1894	8/11/1894	8/11/1894	Offg. 2nd in comd.
	WING OFFICERS.				
23/5/1885	Allenby, Capt. A. H. (p.s.)	23/5/1896	2/4/1887	1/11/1889	Offg. Wing Comdr., Sambalpur.
11/2/1888	Jacks n, Lieut. R. P. (p.s.)	31/7/1889	12/5/1890	17/4/1891	Adjutant, 10th Aug. 1891. Lv. out of India, m.c., 1 yr., 27th Nov. 1897.
16/11/1887	Prince, Lieut. W. C. S. (p.s.)	22/1/1890	20/4/1890	3/5/1891	Q.M. 10th Aug. 1891. Offg. Adjt. and Stn. Staff Offr., Raipur
15/2/1888	Baker, Lieut. H. R.	31/7/1889	1/2/1891	9/9/1892	Stn. Staff Offr., Sambalpur.
25/7/1891	Burn, Lieut. G. C.	11/10/1892	11/10/1892	12/10/1894	Lv. out of India, m.c., 1 yr., 31st July 1897.

(p.s.) Passed for Staff employ.

EXTRACT FROM THE INDIA ARMY LIST, DATED 1ST JULY 1898—*continued.*

First Commission.	Rank, Names, and Corps.	Army Rank.	First Appointment to Regiment.	Present Appointment in Regiment.	Remarks.
		MEDICAL OFFICER.			
31/3/1877	Chatterjie, Surg.-Lieut.-Col. N.	13/3/1897	23/11/1882	23/11/1882	India, m.c., 6 mos., 2nd April 1898.
		ATTACHED.			
9/4/1892	Anderson, Lieut. W. C.	17/9/1894	17/9/1894	17/9/1894	Lv. out of India, m.c., 6 mos., 15th April 1898.
28/9/1895	Plant, Lieut. W. C. T. G. G.	27/1/1897	18/2/1898	18/2/1898	Offg. Wg. Offr.
2/4/1881	Kanga, Surg.-Maj. J. K.	2/4/1893	15/8/1897	2/4/1898	Offg. Medical charge.

Date of entering Service.	Names.	Date of Commission as			Remarks.
		Jemadar.	Subadar.	Subadar-Major.	
		SUBADAR-MAJOR.			
20/5/1867	Jagannath Singh, *Bahadur*	16/4/1886	11/4/1888	6/7/1895	...
		SUBADARS.			
10/3/1869	Kadir Muhiyuddin	30/6/1887	26/4/1890	...	...
1/5/1874	Sayyid Umar (14).	20/3/1888	26/12/1893	...	...
16/6/1874	Viraraghavulu (14)	26/4/1890	6/7/1895	...	Sambalpur.
17/6/1875	Sikandar Khan .	1/3/1891	1/11/1895	...	Sambalpur.
23/3/1877	Abdur Razzak (14) (15)	1/1/1892	1/1/1897	...	Sambalpur.
25/9/1879	Adinarayadu .	26/12/1893	1/3/1897	...	...
6/9/1882	Manikkam Mudali (14)	6/7/1895	24/1/1898	...	...
		JEMADARS.			
28/8/1875	Muhammad Ghaus (14)	9/8/1892	...	...	Sambalpur.
27/4/1877	Abdur Rahman Khan	2/6/1893	...	...	Sambalpur.
9/4/1880	Sayyid Muhiyud-din (14)	24/1/1895	...	...	...
25/2/1880	Ranganayakulu .	25/5/1895	...	...	Sambalpur.
10/10/1882	Azhagarsami .	1/11/1895	...	...	Sambalpur.
22/6/1880	Amir Beg . .	1/1/1897	...	...	...
10/8/1881	Venkatasami .	1/3/1897	...	...	...
...	Ponnusami . .	24/1/1898	...	...	...

(14) Certificate of Qualification in Musketry.
(15) Qualified as Military Surveyor.

PART II

ESTABLISHMENT AND RECRUITING

II

ALTERATIONS IN THE ESTABLISHMENT OF THE REGIMENT, AND THE MEANS BY WHICH IT HAS BEEN RECRUITED

THE establishment of the regiment when first raised was 1776 fixed at two grenadier and eight battalion companies,[1] each company consisting of 1 subadar, 1 jemadar, 5 havildars, 5 naicks, 1 drum, 1 fife to each grenadier company, 1 puckally, and 65 sepoys. The complement of European commissioned and non-commissioned officers consisted of 1 captain,[2] 4 lieutenants, 6 ensigns, and 10 sergeants—total of all ranks, 823.

The tallest and finest men were selected for the grenadier companies, who were distinguished from the rest by tall gaily-decorated turbans and smarter uniforms. These companies held the place of honour, namely, the flanks when in line, and the front and rear when in column for attack. The names of all men who in any way distinguished themselves were ordered to be taken down by the officers, and they were to be put into the grenadier companies, so that these companies should be composed of "real, brave, good men." The grenadier companies of the 13th Carnatic

[1] "Each company was commanded by an European officer, under whom were the native commissioned officers, Subidars and Jemidars (or captains and subalterns) . . . Native officers only could sit on Courts-Martials on the sepoys for transgressing the Articles of War, which were translated into the Persian, Moorish, and Malabar languages, and read to each corps monthly. At General Courts-Martials a British officer superintended the proceedings as Judge Advocate."—Captain C. GOLD.

[2] "Captains were selected to command native infantry corps, and were almost invariably chosen from their reputation as sepoy officers; that is, officers who united to all the military qualifications of a soldier, a particular knowledge of the prejudices, habits, and characters of the men whom they were appointed to command."—MALCOLM.

Battalion were frequently detached from the regiment on special military service, and were usually the companies selected for the actual assault on a fortress when the services of the regiment were required for this duty.

1777 On the 17th January 1777 the strength of the native ranks was increased by two battalion companies being added, and by the grenadier companies being increased by 1 jemadar, 1 havildar, 1 naick, and 25 sepoys—total of all ranks, 1039. These two extra companies would appear to have been used as light infantry, and to have been lightly equipped for skirmishing and other purposes. Men of small but active build were usually posted to these companies. The drill staff was composed of 1 drill sergeant, 2 drill havildars, and 3 drill naicks.

1782 After the formation of six new regiments in 1782, some of the supernumerary levies were drafted into the regiment to complete the strength, and orders were given for the formation of several strong independent companies at Trichinopoly, Madura, and Palamcottah, with a view to using them as depôts from which to recruit the native army.

During the campaigns of 1782–84, the recruits were obtained from amongst the "Moormen, Rajahpoots, Gentoos, Malabars, and Pariahs," and although it was considered injudicious to admit the latter into battalions with men of reputable castes, it was admitted that this mixture was extremely beneficial, as "it stimulated by emulation, and restrained all dangerous confederacies, which could not escape the jealousy of contending sects." The boys who carried the knapsacks of the European soldiers on the march, and who cooked for them in camp, were frequently enlisted as sepoys. Captain Innes-Monro gives the following account of them :—

"These cook boys are amazingly attached to their masters, and will keep close to their heels in the midst of the greatest dangers. When they grow up they make the best sepoys, for all of them speak English well."

On the 5th November 1783 the establishment of European officers for the regiment was fixed at 1 captain, 1 captain-lieutenant, 2 lieutenants, and 7 ensigns; the captain-lieutenant having the rank of captain in the army, but not being eligible for either of the regimental staff appointments. 1783

On the 12th July 1784 the number of European sergeants with the regiment was reduced to three, namely, the sergeant-major, the quartermaster sergeant, and the drill sergeant. 1784

In October of the same year the strength of the regiment was reduced to nine companies, one of which was to be composed of grenadiers, and the establishment was fixed at 1 captain, 1 captain-lieutenant, 2 lieutenants, 6 ensigns, 3 sergeants, 1 native commandant, 8 subadars, 9 jemadars, 36 havildars, 36 naicks, 9 drummers, 5 fifers, 9 puckallies, and 495 sepoys.

On the 23rd August 1785 the companies of the regiment were increased again to ten, and the rank and file increased by 150. At the same time the appointment of native commandant was abolished. This native officer had always taken post in front of the regiment with the Captain-Commandant. 1785

On the 22nd October 1785 it was notified to the army that "the Honourable Company having been pleased to allow one son of each sepoy who has been killed, or died in the service, to receive the pay of his deceased father, the Commissary of Musters is therefore to allow two boys, the sons of sepoys so killed or dead, to be inserted in the muster-roll of each company, amounting altogether, with the effectives, to the fixed establishment; and a remark is to be made opposite to the name of each boy, so that he may not be mistaken for an effective; but when arrived at the age of puberty, such boy, if able-bodied, must enlist, or be struck off the roll"; and thus the establishment of recruit boys[1] commenced.

About the end of 1785, a native doctor was attached to the regiment for the first time, who was paid partly by Government and partly by the native ranks.

1786 On the 20th May 1786 the number of companies in the regiment was reduced from ten to eight, two of which were to be grenadier companies, and the establishment was again revised as follows:—1 captain, 4 lieutenants, 4 ensigns, 8 sergeants, 8 subadars, 8 jemadars, 32 havildars, 32 naicks, 8 drummers, 8 fifers, 544 sepoys, and 8 puckallies. The staff consisted of an adjutant, an assistant-surgeon, and a native doctor in addition to the above.

In this year the enlistment of weavers was directed to be discontinued.

1787 "In lieu of the former establishment of two boys per company, the sons of relatives of old soldiers of the several ranks were allowed for each battalion, exclusive of the regular number of sepoys. These boys were to be admissible at the age of eleven, provided they were healthy and well-limbed, and were to be transferred to the ranks as soon as they became fit." This arrangement came into force on the 1st November 1787.

1790 Early in 1790 the regiment was augmented by two companies, and the establishment raised to ten. It was ordered at the same time "that the native officers, promoted in consequence, should not be confirmed until they furnished a certain number of approved recruits, namely, a jemadar for promotion to subadar, 38; a havildar for promotion to jemadar, 18; and a

[1] These boys were then called *Sepoy Recruits*, and received about two and three-quarter rupees per mensem, and, "being formed to the service from infancy, made excellent officers."—Captain C. GOLD.

naick for promotion to havildar, 12." In consequence of this augmentation the establishment of recruit boys was raised from 40 to 50.

In June 1792 the two additional companies which had been raised at the commencement of the war were reduced. 1792

"In November the establishment of subalterns in each battalion was fixed at 7 lieutenants and 2 ensigns, inclusive of the adjutant, with the view of preventing any variation in the number of each grade."

On the 6th June 1793 a draft was sent from this, as well as from other regiments, to form the nucleus of the 31st Battalion, which was being raised in Trichinopoly. 1793

Colonel Welsh, in his *Reminiscences* (edited in 1830), gives the following description of the different classes from which the Madras Infantry was enlisted in 1793:—

"The native army is composed of five distinct castes, or classes of men, differing most essentially in manners, in religion, and in customs; who never unite, even at a meal, or in marriage; the discipline and harmony which have ever distinguished those native forces are truly wonderful. The more especially, when the bigotry of one class, and the superstitious prejudices of three others, are taken into consideration. . . . First, the Mussulman, of whom at least one-third of the army is composed. This class is again subdivided into four particular sects, namely, the Sheik, the Syed, the Mogul, and the Puthaun, or Pattan, as they are usually called. They are generally brave, enterprising, and intelligent; and, upon the whole, being free from religious prejudices, make excellent soldiers. Second, the Rajahpoot, or descendants of the ancient rajahs, the highest caste of Hindoos; a race not very numerous, but extremely scrupulous; and, when their prejudices are humoured, the bravest and most devoted soldiers, far surpassing all the other natives, in a romantic, but sometimes mistaken notion of honour. Third, the Telinga, or Gentoo, a race of Hindoos generally remarkable for mildness of disposition and cleanliness of person; obedient and faithful, but not very intelligent or enterprising soldiers. Fourth, the Tamoul, or Malabar, similar to the former. Fifth, the Pariah, or Dhére, as they are called in the army. The latter class, poor Chowry Mootoo, brave, active, and attached as they were to their officers and the service, with a few European failings, such as dram-drinking, and eating unclean meats, etc., have of late years been excluded from the line, in order the more fully to conciliate the higher classes; who, however they may differ from each other in many points, are all united in considering any mixture with these as a contamination. They are now enlisted only in the pioneers, and as artillery and tent lascars."

1794 In order to take the places of the men of the 13th Battalion who had volunteered for the expedition against the French in Mauritius in 1794, a large number of recruits were enlisted at regimental headquarters. The expedition having been abandoned, the volunteers returned to their regiment, which was now considerably over strength. On the 12th August 1794 all the newly enlisted recruits were sent to Tanjore as one of the drafts for the formation of the 36th Battalion (the present 26th Madras Infantry).

1795 In 1795, owing to the small pay of the sepoy and the dearness of food, great difficulty was experienced in procuring good recruits, and battalions were kept up to strength by accepting under-sized men, and men of low caste.

1796 On the reorganisation of the native army in July 1796, the 3rd and 13th Battalions having become the 1st and 2nd Battalions of the 3rd Regiment respectively, the establishment of the regiment, composed of two battalions of eight companies each, was fixed at 1 colonel, 2 lieutenant-colonels, 2 majors, 7 captains, 1 captain-lieutenant, 22 lieutenants, 10 ensigns, 20 subadars, 20 jemadars, 100 havildars, 100 naicks, 40 drummers and fifers, 1800 privates, 160 recruit boys, and 20 puckallies, exclusive of a staff consisting of 2 adjutants, 1 paymaster, 1 surgeon, 2 assistant surgeons, 2 sergeant-majors, 2 quartermaster sergeants, 2 native adjutants, 2 native doctors, 2 drum-majors, 2 fife-majors, 2 drill havildars, and 2 drill naicks.

This was the first occasion on which the regiment was supplied with special permanent European medical attendance.

Under this new organization the left wing of the 22nd Battalion (one of the battalions which had been reduced), being formed into three companies, was incorporated into the 2nd Battalion, 3rd Regiment.

1798 In October 1798 the 2nd Battalion, 3rd Regiment, contributed 2 subadars, 2 jemadars, 8 havildars, 8 naicks, 2 drummers, and 200 privates for the formation of the present 27th Madras Infantry, which was raised at Trichinopoly, and brought on the establishment as the 1st Battalion, 14th Regiment.

1800 "On the 25th June 1800 the appointment of 'adjutant and quartermaster' was abolished, as the duties connected with the stores were performed by a sergeant." Thus, from this time, adjutants had only their own departmental duties to perform.

On the 11th July 1800 it was ordered that the native adjutant should be chosen from among the jemadars only.

In July 1800 the regiment furnished a draft of 2 subadars, 1 jemadar, 7 havildars, 7 naicks, and 1 drummer for the formation of the 2nd Battalion, 18th Regiment (late 36th Madras Native Infantry), at Nundydroog.

About this time recruits were chiefly obtained in the Carnatic, and although the men were small, they possessed valuable qualities, thus described: "Although the men obtained in the more southern countries are much inferior to the northern recruits in caste, size, and appearance, they are nevertheless hardy and thrifty, and, being less subject to local attachments, and little encumbered with religious habits or prejudices, to interfere with the regular performance of their duty, are found to stand the pressure of military hardships with much fortitude, and to manifest at all times a firm adherence to the service"—a description of which their descendants may well feel proud.[1]

[1] Captain R. Gold (writing about 1791) says:—"Though the Coast sepoys are frequently of small stature, they have a very soldierlike appearance; and from the high state of perfection their officers have brought them to, by a strict discipline and a study to make their lives comfortable, they appear really attached to the service."

1806 On the 1st January 1806, two dressers paid by Government were allowed to be borne on the strength of the regiment, after which, stoppages from the pay of the sepoys for medical aid to themselves and their connections ceased. Up to this time the dressers had been borne upon the rolls of the regiment as private soldiers only.

1807 On the 1st January 1807 the system of territorial recruiting was introduced, but not carried out, however, until the 1st October of the same year, when the invalid companies attached to the 2nd Battalion, 3rd Regiment, stationed at Palamcottah and Tinnevelly, recruited for this battalion. It was also decided that the families of the battalion, if sent on service, should reside under the protection of the invalid companies, under a European officer if possible.

1808 On the 1st March 1808, 1 jemadar, 2 havildars, and 7 naicks were transferred to the 2nd Battalion from the 1st Battalion of the regiment.

1809 During June 1809 the establishment of privates on the strength of the battalion was reduced from 900 to 800.

1812 On the 1st October 1812 the establishment of the battalion was reduced to 780 privates, all men in excess of that number being returned as supernumeraries until absorbed.

On the 15th November the practice of distinguishing companies by letters was introduced. The grenadier companies were excepted from this regulation. It was ordered also that every man should, on enlistment, have two numbers assigned to him, namely, a general and a company number.

1813 On the 23rd February 1813 the establishment of the recruit and pension boys was fixed at 30 and 40 respectively. The recruit boys were not to be taken under the age of eleven years, but the pension boys were admissible at any age under eleven years. Both, however, were to be relatives or connections of old soldiers of the several ranks.

On the 10th August 1813 a 1st and 2nd dresser were attached to each battalion of the regiment, the former upon a monthly pay of ten, and the latter on a monthly pay of seven star pagodas.

1814 The recruiting of the battalion by means of the invalid companies attached to it, which was introduced in 1807, was discontinued from the 1st January 1814.

On the 26th April 1814 the establishment of the battalion was reduced from 780 to 750 sepoys.

1816 In order to provide recruits for native battalions employed beyond the frontier, a depôt was established at Ellore in May 1816 under a European officer.

1817 On the 7th May 1817 the establishment of the battalion was fixed at 900 privates, but the order being shortly afterwards cancelled, all men in excess of the former strength were ordered to be borne upon the returns as supernumeraries.

In July 1817 orders were received for one of the subalterns to be appointed interpreter of the Hindustani language to the battalion, in conjunction with the appointment of quartermaster.

"A considerable number of Pariahs and other low caste men were serving in the Madras infantry battalions between the years 1776 and 1817, from which it would appear that the sepoys who fought through the Mysore and Mahratta wars, and gained the series of victories from Chengamah in 1767 to Mahidpore in 1817, were largely recruited from the classes now found in considerable numbers in the Sappers and Miners. But it has been suggested that this does not show conclusively the general composition of the service between these years, since low caste men probably volunteered for foreign service with greater alacrity than those more restrained by prejudice."

1818 At the conclusion of the second Mahratta War, many Hindustani recruits were drafted into the regiment, who were probably enlisted by the Madras Regiments then serving out of their Presidency.

The following is a description of the composition of the Madras Native Infantry in 1818:—

"The Native Infantry of Madras is generally composed of Mahomedans and Hindoos of good caste; at its first establishment none were enlisted but men of high military tribes. In the progress of time a considerable change took place, and natives of every description were enrolled in the service. Though some corps, that were almost entirely formed of the lowest and most despicable races of men, obtained considerable reputation, it was feared that encouragement might produce disgust, and particularly when they gained, as they frequently did, the rank of officers. Orders were, in consequence, given to recruit from none but the most respectable classes of society; and many consider the regular and orderly behaviour of these men as one of the benefits which have resulted from this system.

"The infantry sepoy of Madras is rather a small man, but he is of an active make, and capable of undergoing great fatigue upon a very slender diet. We find no man arrive at greater precision in all his military exercises; his moderation, his sobriety, his patience, give him a steadiness that is almost unknown to Europeans; but although there exists in this body of men a fitness to attain mechanical perfection as soldiers, there are no men whose mind it is of more consequence to study. The most marked general feature of the character of the natives of India, is a proneness to obedience, accompanied by a great susceptibility of good or bad usage; and there are few in that country who are more imbued with these feelings than the Madras sepoy.

1819 On the 2nd February 1819 the rank of subadar-major was introduced into the native army with the sanction of the Court of Directors. At the same time the grade of colour-havildar was introduced, one colour-havildar being allowed to each company.

The 2nd Battalion, 3rd Regiment, furnished a draft of 1 jemadar, 1 havildar, 6 naicks, 5 privates, (all of whom received a step in rank), 1 drummer, and 10 privates, for the formation of the 2nd Extra Battalion raised in Trichinopoly in 1819. Only men of good caste and character, and who were active, intelligent, and well-formed soldiers, were ordered to be sent.

The grade of captain-lieutenant was abolished on the 4th March 1819.

The regiment also furnished a draft in the same year of 1 naick and 1 private (who received a step in rank) to the Golundaze Battalion. A draft of 1 havildar, 4 naicks, 3 privates (all of whom obtained the next higher rank), 1 drummer, and 10 privates, was sent to the 2nd Battalion, 25th Regiment, at Palamcottah.

1820 The regiment was allowed a bazaar from the 1st January 1820, which was placed under the immediate control of the Commanding Officer. Ground for this purpose was allotted near the lines. The number and pay of the bazaar establishment was also fixed.

1821 The infantry recruiting depôt at Ellore was abolished on the 28th February 1821. About this time the enlistment of men of low caste was discouraged at headquarters, probably with a view to conciliate the higher classes in the native ranks. Consequently the former excellent class was gradually diminished, and "recourse had to the fairer, taller, and more cleanly men of higher caste." In the 1st Battalion of the regiment (the present 3rd Palamcottah Light Infantry) there were at this time not less than 96 Christians and Pariahs in the regiment.

1822 In September 1822 native pensioners were allowed to choose their places of residence, instead of being compelled to reside with the invalid companies.

Subalterns were promoted brevet-captains after fifteen years' service.

1823 In February 1823 officers commanding battalions were directed to be particular in recruiting, and were forbidden to entertain any man under 5 feet 5 inches in height, or above twenty years of age.

By an order dated Fort St. George, 3rd June 1823, the Honourable Governor in Council was "pleased to direct that in future one petty maistry of bamboo coolies shall be attached to each battalion of native infantry when marching from one station to another, or when employed in the field."

1824 On the 6th May 1824, when the 2nd Battalion, 3rd Regiment, became the 13th Regiment once more, the establishment of officers in the regiment was fixed at 1 colonel, 1 lieutenant-colonel, 1 major, 5 captains, 10 lieutenants, and 5 ensigns. The complement of British officers was increased, owing to the need for their services being emphasised by the fourth Mahratta or Pindaree War, and to the great wastage which took place during the campaign.

During this year orders were issued for the recruiting for each Presidency to be confined to its own limits.

On the 6th August of the same year the strength of the regiment was raised to 800 privates, and shortly afterwards to 850 privates. On the 13th September 1824 the strength was again increased to 900 privates, and in January of the following year to 950 privates.

1825

1826 On the 16th January 1826 the regiment furnished a draft of 1 havildar, 3 naicks as havildars, 3 privates as naicks, 1 drummer and 3 privates, for the 3rd Extra Regiment.

On the 28th January 1826 a General Order was published directing sepoys to be vaccinated.

On the 25th August 1826 the establishment was reduced to 800 rank and file.

1827 On the 20th March 1827 the Governor in Council published orders for the formation of an establishment of "native medical pupils," and, after qualifying, these men were posted to regiments in the grade of second native dresser.

1829 On the 15th January 1829 the standard height for recruits was fixed at 5 feet 6 inches, and for regimental lascars 5 feet 4 inches. All recruits to be between the ages of sixteen and twenty-two years.

In obedience to orders from the Honourable the Court of Directors, the Right Honourable the Governor-General in Council issued orders from Fort William on the 5th May 1829 for the reduction of the establishment of the regiment by two companies with

2 lieutenants and 1 ensign. The strength of each company to be 80 privates. At the same time the establishment of European officers was fixed at 1 colonel, 1 lieutenant-colonel, 1 major, 5 captains, 8 lieutenants, and 4 ensigns. These alterations had effect from the 5th June 1829.

On the 23rd April 1830 orders were promulgated for the establishment of regimental schools in each corps. Two schools were to be formed, one Mussulman and one Hindoo, and two masters were to be appointed. Two of the regimental lascars were to be employed as under-masters or ushers. The schools were to be managed by a committee consisting of the Commanding Officer as president, with the interpreter and another officer as members. The interpreter was to act as secretary. Each member of the committee was to superintend one of the schools. All regimental boys were required to attend, and the schools were also to be free to all boys whose fathers or relatives were, or had been, in the ranks. Such men of the regiment, particularly recruits, as wished to attend, were to be permitted to do so.

1830

On the 10th August 1830 the Governor-General was pleased to resolve that a moonshee should be appointed to each regiment of native infantry, the allowance for interpreting heretofore drawn by quartermasters of native corps being discontinued. The services of the moonshee were to be available for the instruction of any officer of the corps on payment of five rupees per mensem.

On the 9th August 1836 the appointment of regimental vakeel was discontinued.

1836

On the 31st August of the same year the office of regimental moonshee was abolished, and the allowance of thirty rupees per mensem reverted to the interpreter of the regiment for the purpose of providing a moonshee, on a salary of that amount, agreeably to the practice which obtained prior to the promulgation of the General Order of the 10th August 1830. The moonshee was to be borne on the establishment of the interpreter and quartermaster.

Between the years 1802 and 1836, 15 "Christian Pariahs," 9 Indo-Britons, and 2 Europeans were enlisted or transferred from other regiments for the band and drums.

G.O.G., dated 4th October 1836, directed "the whole conduct of the pay department of corps to rest in future with officers commanding companies and heads of departments respectively, who will be severally responsible for all matters connected with the drawing and issuing of all pay or other money due to those under their charge. Regimental quartermasters are accordingly relieved from the duty of collecting and examining regimental abstracts, as well as of attending at the pay office to receive this amount."

1837 On the 17th February 1837 bhistis were substituted for puckallies.

1838 On the 1st May 1838, "in assimilation to the practice in Bengal, the Governor in Council was pleased to extend, henceforth, the benefit of medical aid in garrison, and in the field, to all classes of followers permanently or temporarily receiving pay from Government."

On the 6th September 1838 the established strength of each company was fixed at 90 privates. Owing to this increase in strength of ten men per company, officers commanding regiments were directed to detach small recruiting parties. Men of 5 feet 4 inches were permitted to be enlisted, provided they were of active make, athletic, and not under sixteen years of age.

1839 On the 13th July 1839 the strength of the band was increased from 1 havildar and 10 private musicians to 1 havildar and 14 musicians.

G.O.G., dated 16th August 1839, authorised the formation of the 7th battalion company, the established strength of each company in the regiment to be 1 subadar, 1 jemadar, 6 havildars, 6 naicks, 2 drummers, 3 lance-naicks, 100 privates, 1 lascar, and 1 bhisti.

On the 3rd October 1840 an addition of 1 lieutenant and 1 ensign to each regiment was sanctioned. 1840

As authorised by G.O.G., dated 18th February 1842, an additional 8th battalion company was formed of the same strength as the others. 1842

On the 20th September 1842 a tenth lieutenant was added to the regiment.

On the 27th January 1843 the established strength of each company was 90 privates. 1843

The complement of non-commissioned officers of native regiments was reduced by 1 havildar and 1 naick in February 1843, leaving the strength of each company at 1 subadar, 1 jemadar, 5 havildars, 5 naicks, 2 drummers, and 90 sepoys (Fort St. George Orders of 28th February, and Fort William Orders of 17th February).

On the 4th February 1845 a General Order of Fort St. George authorised an additional or sixth captain to the regiment. 1845

As a specimen of the class of recruits obtained over fifty years ago, the following recruits were enlisted at Secunderabad and Chicacole during the year 1845:—6 Muhammedans, 10 Shaiks, 4 Syeds, 1 Shepherd caste, 2 Christians, 1 Indo-Briton, 7 Telingas, 3 Buljawars, 1 Malabar Washerman, 3 Pariahs, 5 Rajputs, 1 Brahmin, 2 Vellauls, 1 Gooraul, 1 Agumoodya, 1 Gomulhausy, 1 Padiyachee, 1 Kurruum. During the same year 1 Gentoo, 1 Malabar, 1 Indo-Briton, and 1 Vellunlum were transferred from other corps.

In August 1846 a recruiting party under Havildar Shaik Emam was sent to Vellore, where 9 Muhammedans and 1 Malabar were enlisted. The following recruits were enlisted at Secunderabad:—3 Syed Muham- 1846 medans, 1 Gentoo, 5 Indo-Britons, 1 Rajput Shepherd, 14 Shaik Muhammedans, 1 Malabar Oilmonger, 3 Telingas, 13 Pathan Muhammedans, 6 Buljawars, 1 Goldsmith Tamil, 1 Agumoodya caste Tamil, 2 Malabars, 4 Rajputs, 1 Vulloowar, 1 Padiyachee, 1 Telinga Shepherd, 3 Brahmini Rajputs, 1 Gentoo Mootrass, 1 Rajput Chutri, 2 Gentoos, 1 native Christian.

1847 G.O.G., dated 11th February 1847, fixed the established strength of the regiment at 700 privates, leaving the strength of each company at 70 privates.

1856 An additional captain and lieutenant were added to the establishment of every native infantry regiment (G.O.G., 11th November 1856), with effect from 23rd November 1856.

1857 On the 20th November 1857, in consequence of the mutiny of the Bengal Army, all Madras Infantry Regiments were increased to the strength of 1000 privates each, and an extra havildar and naick per company were authorised, owing to the strength of each company being augmented by 30 privates (G.O.G., No. 367, dated Fort St. George, 20th November 1857).

1858 An addition of 1 lance-naick per company of native infantry was sanctioned (G.O.C.C., No. 22, dated 26th February 1858).

The following recruits were transferred from the recruiting depot, Ellore, on the 30th July 1858:—39 Gentoos, 30 Muhammedans, 4 Gentoo Coomers, 1 Gentoo Barber, 6 Gentoo Dunghers, 10 Pariahs, 6 Gentoo Weavers, 4 Gentoo Dhobies. During the mutiny in Bengal, many recruits were enlisted, and notwithstanding the increase in the strength of every Madras Regiment, the large demand was met without difficulty.

On the 31st August 1858, one hundred privates were sent from the recruiting depôt at Chicacole to Maulmain to join the regiment, which had fallen considerably under strength. These recruiting depôts had been again established, and enlisted men for the different regiments of the Madras Army; the one at Chicacole despatching 1075 recruits to various native infantry regiments on the above date. The following is the proportion in which the recruits were enlisted at the depôt:—one-fourth Tamilians from the Carnatic and Ceded Districts, one-fourth Telingas from the Northern Circars, one-fourth Muhammedans, and one-fourth natives of lower castes, or without recognised castes. (These recruits all wore red facings on their coats.) At Maulmain, 2 Muhammedans, 2 Gentoos, 2 Vellauls, 1 Christian, and 1 Malabar were enlisted.

1859 On the 19th January 1859 one naick and one private were transferred from the regiment, each receiving a step in rank, to the 4th Extra Battalion, which was being raised at Chicacole.

1860 On the 1st January 1860 the establishment was reduced to 700 privates per regiment. The following is an extract from G.O.G., No. 2, dated 3rd January 1860, Fort St. George:—

"The Honourable the Governor in Council directs that, from the 1st instant, each regiment of native infantry of the Madras Army be reduced to 50 havildars, 50 naicks, and 700 privates. All men in excess of this establishment are to be borne on the returns as supernumeraries until absorbed by casualties, or otherwise disposed of; until the numbers in the non-commissioned ranks, however, are reduced in each regiment to the new standard, one promotion will be made for every two vacancies. The strength of each company will be as follows:—1 subadar, 1 jemadar, 5 havildars, 5 naicks, 2 drummers or buglers, 70 privates. All men supernumerary to the reduced establishment of three years' service, or any stated period, are to be paid up and discharged, with a gratuity of one month's pay and half batta (7 rupees) for each year of service."

By G.O.C.C., dated 17th June 1860, the addition of one lance-naick per company, sanctioned on the 26th February 1858, was cancelled.

In order to provide for the duties of camp colourmen, heretofore performed by the regimental lascars on the line of march, one sepoy from each company was appointed camp colourman. This order had effect from the 30th March 1860.

On the 4th April 1860, 1 subadar, 5 havildars, 2 naicks, and 21 privates were transferred to the regiment from the 2nd Extra Regiment, which had been disbanded.

The following is an extract from the G.O.C.C., No. 108, dated 26th October 1860, reducing the strength of the regiment :—

> "The Commander-in-Chief directs it to be notified that, with the sanction of the Home Authorities, Government has decided on the reduction of regiments of native infantry to the establishment of 600 privates, each formed into eight companies. . . . The establishment of the native ranks of regiments of native infantry will hereafter be as follows : 8 subadars, 8 jemadars, 41 havildars, 40 naicks, 16 drummers, 600 privates, 24 recruit boys, 32 pension boys, 8 puckallies, 1 moonshee, 1 armourer-maistry, 1 assistant armourer, 1 chuckler, 1 bellows boy, 1 chowdry, 2 peons, 1 2nd tindal, 10 lascars, and 2 toties."

1861 By G.O.G., No. 322, dated 20th September 1861, the number of regimental lascars of native infantry regiments was fixed from 1st October 1861 at one 2nd tindal and one lascar for each company.

In 1861 a recruiting party was sent to Ellore and Trichinopoly. Sixteen recruits were enlisted.

1862 On the 6th March 1862, 1 subadar, 2 jemadars, 11 havildars, 12 naicks, 2 drummers, 68 privates, and 1 recruit boy were transferred to the regiment from the 45th Native Infantry, which had been reduced. On the 31st May of the same year, 1 drummer, 58 privates, and 6 recruit and pension boys were transferred from the 52nd Native Infantry, which had also been disbanded.

1864 On the 11th June 1864, 1 havildar, 1 naick, and 12 privates were transferred from the 18th Native Infantry, which had been disbanded in Cannanore, in consequence of the robbery from the cash chest in charge of a guard of that regiment.

On the 1st July 1864, 1 subadar, 2 havildars, 2 drummers, 7 naicks, and 55 privates were transferred from the 42nd Native Infantry; also 1 puckally, 1 2nd tindal, and 4 recruit and pension boys.

"On the 7th October 1864 the grenadier and light companies of regiments of native infantry were abolished, in assimilation with the change introduced into the British infantry." The remaining companies were ordered to be numbered from 1 to 8, and were ordered to stand on parade habitually according to seniority of officers commanding companies, from flank to centre.

1865

By G.O.G., No. 377, dated 24th October 1865, "the system of organization then in force in the other Presidencies was ordered to be introduced in Madras from the 1st November. The establishment of European officers was fixed at a commandant, senior wing commandant, junior wing commandant, adjutant, quartermaster, and one doing-duty officer." Every native regiment was thus nominally reorganised on the "irregular" basis, a second doing-duty officer being subsequently posted to each regiment.

1866

By G.O.G., No. 312, dated 14th August 1866, the senior wing commandant was designated "second in command and wing officer," and the junior wing commandant was styled "wing officer." On the 17th September 1866 the designation of "doing-duty officer" was changed to that of "wing subaltern," the ordinals first and second being still retained.

In order to assist the native army in maintaining conservancy in the hutting lines, one cart and eight toties per regiment of infantry was sanctioned by G.O.C.C., No. 104, dated 12th October 1866.

About the end of 1866 a recruiting party, under the command of Subadar Sayyid Monadeen, was sent to Vellore, where 17 Muhammedans, 1 Rajput, and 3 Gentoos were enlisted.

During 1866 the following recruits were enlisted at Cannanore:—1 Tamul, 8 Gentoos, 24 Muhammedans, 1 Christian, 1 Indo-Briton, 1 Rajput, and 1 Malabar.

1867 During the following year, Subadar Sayyid Monadeen enlisted 8 Muhammedans, 1 Malabar, and 2 Gentoos at Vellore; and 20 Muhammedans, 1 Indo-Briton, 10 Gentoos, 1 Rajput, 5 Vellorans, 1 Buljawar, 1 Christian, and 1 Tamul were enlisted at Cannanore.

1870 The regiment being about to proceed to Hong Kong for garrison duty, privates from the under-mentioned regiments volunteered for service with the regiment, and were transferred on the 9th September 1870:—

2nd Native Infantry	5	Privates	28th Native Infantry	9	Privates
3rd Light Infantry	9	„	30th „	3	„
8th Native Infantry	11	„	31st Light Infantry	4	„
16th „	37	„	34th „	11	„
17th „	9	„	35th Native Infantry	1	„
23rd Light Infantry	7	„	39th „	6	„
25th Native Infantry	35	„			
27th „	4	„	Total	151	Privates

1871 From the 1st January 1871, in place of being designated by numbers, the companies of native infantry were to be henceforth distinguished in each regiment by letters from A to H, for all purposes of interior economy.

1877 By G.O.G., No. 31, of 1877, the British officers attached to native regiments were, except in the case of the Commandant, placed in two classes as respects designation, wing officers, including the second in command, being designated "wing commanders," and the remainder, including the adjutant and quarter-master, being styled "wing officers." The posts of adjutant and quartermaster were to be filled by one of the wing officers.

1878 On the 2nd May 1878, Sir Neville Chamberlain, the Commander-in-Chief, intimated "that recruits for the native army are not to be enlisted outside the limits of the Presidency, except with sanction obtained through the Adjutant-General. . . . What the sepoys of the Coast army did in times past, their successors, if judiciously selected, may be expected to do in the future. . . . The sepoy, whether of Bengal, Madras, or Bombay, is just what his British officers make of him; and Sir Neville Chamberlain has now seen enough of this native army to feel assured that the Presidency can produce plenty of good efficient soldiers, provided they be sought for, and, when obtained, be properly trained and well commanded."

In 1878 the minimum height for recruits was fixed at 5 feet 6 inches, and the minimum chest measurement at 32 inches above and 31 inches below the nipple. Age not under sixteen years, and not over twenty-two years.

By S.G.O., No. 16, of 1878, the appointments of Commandant and wing commander were only to be held by officers of or above the rank of captain. The appointments of adjutant and quartermaster were only to be permanently held by subaltern officers, and were to be vacated on promotion of the incumbent to the rank of captain. The captain so displaced to revert to the duties of his appointment as wing officer. Wing officers on promotion to major were not to vacate their regimental appointments.

By S.G.O., No. 38, of 1878, no officer who was not in possession of a certificate of musketry would be confirmed in the appointment of regimental adjutant, until he obtained a certificate of qualification.

S.G.O., No. 123, of 1878, sanctioned school certificates of education being granted to soldiers of the native army under certain conditions. The certificates were to be of four classes, and examinations were to be held half-yearly by the School Committee. In the case of candidates for 1st, 2nd, and 3rd class certificates, sealed examination papers were to be furnished by the Superintendent of Army Schools, on whose recommendation the certificate would be awarded. Fourth class certificates were to be awarded on the recommendation of the School Committee. The attendance of recruits at school was ordered to be compulsory for one year after they had joined, and until they had obtained a 4th class certificate.

1882 The following is an extract from G.O.G., No. 264, of 1882, republishing General Order, No. 210, dated 22nd April 1882, of the Government of India:—"Every native infantry regiment will consist, as at present, of eight companies, each company having the following establishment:—1 subadar, 1 jemadar, 5 havildars, 5 naicks, 2 drummers, and 90 privates, total 104; making a total per regiment of 832 natives of all ranks. The Madras Regiments will retain their extra havildar-majors as at present." Eight regiments of Madras Native Infantry, namely, the 34th to 41st, were reduced in June, in order that this increase might be effected.

Consequent upon the above increase in the strength of native regiments, the establishment of European officers was fixed from 1st July 1882 at eight officers per regiment, the additional officer being appointed in the rank of wing officer.

Thirty-nine privates and 7 recruit and pension boys were transferred to the regiment from the 37th Regiment of Grenadiers on the 1st July 1882.

During the stay of the regiment at Jubbulpur, ten Rajputs were enlisted.

From this year dates the difficulty in obtaining recruits of the proper stamp.

1885 On the 31st July 1885, in view to maintaining the native armies as nearly as possible at their full strength, and to provide a margin for casualties, officers commanding native infantry were authorised to entertain supernumeraries to the extent of 20 sepoys per corps in excess of the establishment of 720 sepoys. This order was cancelled in 1890.

"The number of strangers and hangers-on dependent on the Madras sepoy for support, and thus seriously diminishing his means of subsistence, had long been found to be a great evil, with the view of checking which an Order was issued, to the effect that from and after the 1st January 1885, no more than two adults, or one adult with his or her unmarried daughters and male children under sixteen years of age, should be permitted to live with a soldier in the regimental lines. Relations already living there were not to be interfered with. Full discretion was given to Commanding Officers to make exceptions whenever the necessity for so doing should be clearly established."

"It was notified, at the same time, that no recruit, enlisted after the 1st January 1885, should be permitted to marry, or, if already married when enlisted, his wife should not be allowed to reside in the lines, for three years after his enlistment."

On the 19th October 1886 a recruiting party of one naick and three privates was despatched to Trichinopoly. Seven recruits were enlisted.

1886

"In October 1886 orders were issued for the linking together, in regiments of three battalions each, of the native infantry of the three Presidencies. All enlistments in any battalion after the 30th November of the same year were to be made for all the battalions; that is to say, a recruit enlisted for one battalion of a regiment would be liable to serve in either of the other battalions of the same regiment. It was also laid down that during peace there would not be any change, either in the organization of the regiments, or in the conditions of the sepoy's service. When warned for service, the battalion so warned was to make up its strength to 1000 effective rank and file by transfers of men, enlisted under the new conditions, from the other battalions of the regiment, which were then to be recruited to the extent required to replace the men transferred."

The 13th Madras Infantry was linked to the 20th and 22nd Madras Infantry, and Pallaveram, near Madras, was appointed as the regimental centre under the new territorial distribution (I.A.C., Oct. 1888).

In Oct. 1886 an ambulance transport, consisting of 1 dhooly, 1 field stretcher, and 4 bearers, was sanctioned for every native regiment in cantonments; 2 Lushai dandies with covers for ordinary movements; 1 Lushai dandy, 4 bearers, and 8 field stretchers, when on active service.

On the departure of the regiment from Bellary on field service, on the 7th December 1886, some of the families of the native ranks were despatched to their native villages, but the majority of them remained behind in Bellary, under the charge of Subadar Ramasami.

About 100 men were left behind at the depôt (the majority of them being under transfer to the pension establishment), under the command of Lieutenant B. Holloway, 2nd Madras Lancers.

1887 "When the order regarding linked battalions was issued, the formation of a system of reserves for the native infantry of the three Presidencies was directed at the same time, but did not have effect in Madras until 1st April 1887. The reserves were to be composed of an active and a garrison reserve. The following were the principal features of this measure. The active reserve was to be formed of men of not less than five or more than twelve years' service with the colours, and they were to be liable for field service with either of the battalions linked with their own. The number of men to be drafted from each battalion for this service was limited in the Madras Presidency, for the present, to 160 men. This reserve was to be embodied for training for one month every year. The garrison reserve was to be formed of men pensioned after twenty-one years' service, and of those belonging to the active reserve who had completed a service of twenty-one years. This reserve was to be employed on garrison duty, and not liable to serve beyond the frontier of British India. It was to be embodied for training for one month every alternate year, and the number belonging to it was unlimited."

1888 Paragraph 29 of the Army Reserve Regulations, Sec. IV. Act IV. of 1888, suspended the enrolment in, or transfer to, the garrison reserve for the present.

With a view to the employment of none but soldier clerks for the native army, in supercession of the system under which regimental officers entertained their own writers, the Government of India, on the 13th January 1888, sanctioned certain rules for the selection, promotion, period of appointment, rates of extra duty pay, etc., of soldier clerks. The clerks were to be trained regimentally.

With a view to provide a trained establishment of ward servants for employment as nursing orderlies in hospitals of native troops, the Government of India sanctioned the enlistment, after the 21st December 1888, of two ward servants per battalion of native infantry as privates, to be included in the present fixed establishment of corps. These men were to be enlisted for "general service" in war time, though in peace they were not to be liable to transfer. They were to be granted all the privileges of the soldier, as regards good conduct pay, furlough, and pension, but were never to rise above the rank of private.

The strength of the lascars in the regiment was reduced to one tindal and four lascars on the 7th April 1888, the other public followers being eight puckallies and one chowdry.

On the 20th July 1888 it was notified that as many men of the native army as would volunteer, up to a maximum of one hundred per battalion, would be passed into the active reserve. This superseded all previous Orders on the subject.

1889

On the 1st April 1889 the native ranks of the regiment was composed of the following classes:—

Rajputs	34
Telingas	297
Tamils	51
Other castes	36
Christians	31
Muhammedans	283
Indo-Britons	18
Total	750
Wanting to complete strength	83

On the 11th October 1889 a General Order was published reducing the pension boy establishment from thirty-two to eighteen, and fixing the pay of recruit boys at Rs. 5 and pension boys at Rs. 4 per mensem. In order to give effect to these changes, no further enlistment was made after the 31st December 1889, until the establishment now sanctioned was reached, as the increased rate of pay did not come into force until the present excess of pension boys had been absorbed.

On the 28th January 1890 the Bengal, Madras, and

Bombay Staff Corps, being amalgamated, were called "The Indian Staff Corps."

1890 In April 1890 the 10th, 12th, and 33rd Regiments of Madras Infantry were converted, and their places taken by regiments (recruited from several Burma Military Police Battalions composed chiefly of Punjabis and other races of Northern India) localised in Burma. The native ranks of the above three regiments were either discharged and pensioned, or transferred to other Madras regiments. The 13th Madras Infantry being considerably under strength on its return from foreign service in Burma, received 3 naicks, 7 lance-naicks, 66 privates, and 8 recruit and pension boys from the 10th Madras Infantry; 1 lieutenant, 2 drummers, 1 private drummer, and 5 musicians from the 12th Madras Infantry; and 1 lieutenant from the 33rd Madras Infantry.

On the 9th July it was decided to reduce the numbers of Northern Circar Telingas serving in Madras Infantry Regiments, and Commandants were desired to enlist as few recruits as possible of this caste. Consequently no further recruiting was carried out in the districts in which these Telingas mostly reside.

On the 24th August 1890 a recruiting party, consisting of one naick and four privates, was sent to Vellore. Eleven recruits were obtained.

The number of "Telingas" in the Madras Infantry being largely in excess of all other castes of which these regiments are composed, His Excellency the Commander-in-Chief, with the sanction of Government, directed in General Orders the discontinuance of further recruiting from men of that caste, from the 5th December 1890, and reduced the standard height for infantry recruits under twenty years of age from five feet five inches to five feet four inches. The secret of the number of Telingas in the Madras Army has been assigned to the more backward economical condition of the Telugu country and people, on the supposition that in civilising a country and developing its resources, the local supply of recruits is cut off. The old fighting castes, the Rachwars and the Polygars, are now no longer found in the regiment. "One hundred years ago these castes formed a large proportion of the Telingas in the Coast army, and were especially sought for their bravery and soldierly qualities."

On the localisation of the 32nd Madras Infantry in Burma in January 1891, one naick and thirty-six privates were transferred to the regiment from it with effect from the 1st February.

1891

On the 23rd January 1891 orders were issued that recruit boys of the Telinga caste now on the establishment could be transferred to the ranks, under the regulations then in force, but enlistment as recruit boys, and transfer to that class, were to cease. Orphans or sons of old and specially deserving men of the combatant ranks of the Telinga caste could, however, be enlisted as pension boys, but they were to be discharged on attaining the age of fourteen years.

The training of garrison reservists was discontinued from the 4th December 1891.

A recruiting party of one naick and four privates was sent to Vellore on the 4th October 1891. The party rejoined regimental headquarters on the 1st January 1892, having enlisted during that time twenty-seven Muhammedans and three Rajputs.

The prohibition against the enlistment of Telingas was modified in November 1891 by the following considerations:—

Linga Belijas, including their subdivision the Jangams, and the Buljawars, or more correctly Balyas, were to come under the prohibition, but a section of the Balyas many years ago emigrated southwards and settled in Tamil districts, and adopted Tamil as their home speech. These men, who are known as Kavaries, and sometimes take the title of Nayakhan, were to be eligible for enlistment, as they, by long associations and surroundings in a Tamil country, had practically become, to all intents and purposes, Tamilians, although originally descended from a Telugu caste.

1892 On the 22nd January 1892, 2 naicks, 5 lance-naicks, and 31 privates, who had been transferred from the 30th Madras Infantry on the localisation of that regiment in Burma, joined the regiment at Bangalore, and were taken on the strength from the 1st of the following month.

On the 4th May 1892 a recruiting party, consisting of one naick and four privates, was sent to Madura.

During the time the regiment was stationed in Bangalore, 113 recruits were enlisted at headquarters: of these 72 were Muhammedans, 8 Tamils, 7 Telingas, 13 Christians, 6 Rajputs, 1 Mahratta, and 6 belonged to other castes.

1894 The first January 1894 found the regiment composed of the following classes:—

Muhammedans	336
Tamils	152
Telingas	179
Rajputs and Mahrattas	43
Christians and Other Castes	110
Total	820
Wanting	13
Establishment	833

In 1894 it was ordered that the families of all native officers, full non-commissioned officers, buglers, drummers, and those of one-third of the remainder of the native ranks, including musicians, enlisted after the 6th December 1889, would be allowed to reside in the lines, such families to consist of wives and children only.

In view of the contemplated formation of class

companies in the Madras Army, the enlistment of Brahmins was prohibited after 20th April 1894.

On the 3rd July 1894, the regiment having fallen twenty-seven below strength, a recruiting party, consisting of one havildar and six privates, was detached to the Madura and Tanjore districts to enlist Tamils, and rejoined headquarters on 13th August 1895. Twenty-three recruits were enlisted.

On the 13th August, in accordance with instructions from Army Headquarters, a party, consisting of one jemadar, one havildar, and eight privates, was despatched into Coorg for the purpose of trying to induce Coorgs and Gaudas to take service, but, after working the whole of Coorg, one recruit only, a Gauda, was obtained. According to the report of the native officer in command of the party, the inhabitants of Coorg, although a warlike race, are much too well-to-do to leave their homes for the small pay of a sepoy.

On the 5th September a recruiting party of one naick and four privates was sent to Vellore to enlist Muhammedans. They rejoined headquarters on the 15th May 1895. Twelve recruits were enlisted.

During this year 22 men were transferred to the reserve, 3 to other corps, and 42 became non-effective, while 33 Tamils, 23 Muhammedans, 3 Mahrattas, and 5 "Other Castes" were recruited.

On the 31st December the regimental reserve numbered 61; of these 16 were garrison reservists.

1895

The enlistment of Mahrattas for service in regiments of Madras Infantry was discontinued from the 11th March 1895.

A recruiting party, consisting of one naick and four privates, was transferred from Kurnool to Cuddapah to enlist Muhammedans, but as only two men were enlisted in four months, the party was recalled to regimental headquarters.

On the departure of the regiment to Burma on the 31st October, the 22nd Madras Infantry (one of the linked battalions) carried out the recruiting for the regiment and trained the reservists.

During 1895, 10 men were transferred to the reserve, 4 to other corps, and 67 became non-effective, while 19 Tamils, 25 Muhammedans, 8 "Other Castes," and 1 Rajput were enlisted, and 19 were received as transfers from other corps. During the same period 11 reservists were discharged and 10 were received, which left a strength of 45 active and 14 garrison reservists at the end of the year.

The regiment being under orders to proceed on foreign service on the 31st October 1895, the families of the native ranks were sent from Cannanore in three parties on the 31st October, 1st and 2nd November, 1895. Vellore having been selected as the family head-quarters, a guard, consisting of one havildar and eleven privates, under the command of Subadar Muttunayudu, was sent there on the 31st October. All the recruit and pension boys, with the exception of one, were left behind at the depôt.

By *London Gazette* of the 6th November 1895, the temporary ranks of lieutenant-colonel and major were conferred on Major H. Lawson and Captain G. A. Welman, whilst holding the appointments of regimental Commandant and Second in Command respectively.

1896 On the 1st January 1896 the regiment was composed of the following classes:—

Muhammedans	347
Tamils	184
Telingas	118
Rajputs and Mahrattas	44
Christians	69
Other Castes	46
Wanting	25
Total	833

On the 10th July 1896, the following Command Order was issued by His Excellency the Lieutenant-General Commanding the Forces, Madras Command :—

"M.C.O., No. 342.

"ORGANIZATION—NATIVE ARMY.

"1. In accordance with instructions received from Army Headquarters, the following provisional composition by class half squadrons and class companies of the existing material in Regiments of Madras Cavalry and Infantry is notified for general information, and will have effect from the 1st September 1896.

"2. The following principles will be adhered to in carrying out the measures, as the classes, which under existing Orders are gradually being eliminated, disappear from the ranks.

"(*a*) No class in any regiment should consist of less than two half squadrons or two companies.

"(*b*) There should not be, as a rule, more than three classes in a regiment, and not more than four under any circumstances.

"(*c*) Promotions of both native officers and non-commissioned officers should be strictly confined to squadrons, or companies, of their own class.

"Their future composition on the lines above-mentioned will be fixed hereafter."

. . . .

INFANTRY.

Linked Battalions.	Muham-medans.	Tamils.	Telingas. (*c*)	Mixed Companies.
... ...	...	...	...	...
13th Madras Infantry . .	3 Coys.	1 Coy.	1 Coy.	3 Coys.
20th Madras Infantry . .	3 Coys.	1 Coy.	1 Coy.	3 Coys.
22nd Madras Infantry . .	3 Coys.	2 Coys.	1 Coy.	2 Coys.
Regimental Centre, Pallavaram	...	...	...	...

(*c*) Telingas are under elimination from all corps.

The following return shows the composition of the regiment by caste and countries on the 1st September 1896, the date from which the provisional formation by class companies had effect :—

Caste.	Subadars.	Jemadars.	Havildars.	Naicks.	Drummers.	Privates.	TOTAL.
Muhammedans.							
Of Madras	4	4	21	16	...	295	340
Hindus.							
Rajputs	1	...	5	5	...	26	37
Tamils	2	1	6	3	...	136	148
Telugus	...	1	7	14	...	113	135
Agambadians, Moravers, Cullers .	1	1	...	...	...	17	19
Madras Hindus of other castes . .	...	1	1	1	...	52	55
Mahrattas.	...	...	...	...	...	6	6
Christians.	...	...	1	1	15	50	67
Total	8	8	41	40	15	695	807
Countries.							
Malabar and South Canara . .	...	...	...	1	...	19	20
Northern Circars	...	2	10	14	...	163	189
Central and Southern Carnatic . .	5	4	19	9	7	186	230
Ceded Districts	...	...	4	4	3	80	91
Hyderabad and Mysore . . .	...	...	4	4	2	64	74
Tanjore, Madura, Tinnevelly . .	1	2	4	4	2	130	143
Belgaum, Salem, Coimbatore . .	2	...	...	3	1	47	53
Coorg, Travancore, Cochin . .	...	...	...	1	...	6	7
Total	8	8	41	40	15[1]	695[1]	807

[1] One drummer and twenty-five privates wanting.

COMPOSITION, ETC., OF THE COMPANIES OF THE REGIMENT ON THE 1ST SEPTEMBER 1896.

Company.	Caste.	Strength all Ranks.	Average Height.	Average Weight.	Average Chest Measurement.	Average Age.
A	Tamil . .	104	5′ 7½″	126½ lbs.	33½″	27 years.
B	Telugu . .	103	5′ 8″	126½ ,,	32¾″	31 ,,
C	Muhammedan .	104	5′ 7¾″	128 ,,	34¼″	27 ,,
D	Muhammedan .	103	5′ 7″	124 ,,	33½″	26 ,,
E	Mixed . .	104	5′ 6¾″	130 ,,	33¾″	27 ,,
F	Mixed . .	83	5′ 7″	129 ,,	34¼″	32 ,,
G	Muhammedan .	103	5′ 6½″	125 ,,	33½″	29 ,,
H	Mixed . .	103	5′ 7″	120½ ,,	33¾″	27 ,,

Of the preceding list, 6 subadars, 2 jemadars, 11 havildars, 4 naicks, 5 lance-naicks, and 49 privates had served in the regiment either as recruit or pension boys, and 8 subadars, 5 jemadars, 29 havildars, 26 naicks, 21 lance-naicks, 12 drummers, and 276 privates were the sons of men who had served or are serving in the native army.

During 1896, 4 men were transferred to the reserve, 3 to other corps, and 45 became non-effective; while 13 Tamils, 15 Muhammedans, and 2 Rajputs were enlisted, and 6 were received as transfers from other corps.

1897

On the 16th February 1897 a scheme of recruiting for regiments of Madras Infantry, to be carried out under the control of British officers, was sanctioned as a tentative measure, and was adopted at once. A certain area was assigned to each group of linked battalions as a recruiting ground, and a selected officer from one of the battalions of the group was detailed to superintend all recruiting operations within that area.

Under the above scheme the civil districts of Ganjam, Vizagapatam, Godaveri, Kistna, and Jeypore were assigned to the 13th, 20th, and 22nd Madras Infantry, and Lieutenant J. S. Hodding, 20th Madras Infantry, was appointed District Recruiting Officer.

The following are the names of the towns from which the greater portion of the recruits referred to in the table of "Classes from which the Regiment has been Recruited since 1845," were obtained:—

Cannanore	287 Recruits.	Palamcottah	118 Recruits.
Trichinopoly	213 "	Vellore	115 "
Madras	169 "	Cuddapah	107 "
Bellary	157 "	Ellore	89 "
Secunderabad	133 "	Vizagapatam	82 "
Bangalore	131 "	Maulmain	42 "
Chicacole	122 "	Tanjore	32 "

1898 In August 1898 the ranks of Surgeon-Lieutenant-Colonel, Surgeon-Major, Surgeon-Captain, and Surgeon-Lieutenant of Officers of the Indian Medical Service, were changed to Lieutenant-Colonel, Major, Captain, and Lieutenant respectively.

PART III

DRESS, CLOTHING, AND NECESSARIES

III

ALTERATIONS IN THE CLOTHING OF THE REGIMENT

"The idea of introducing uniform dress for sepoys arose partly from a desire to improve their appearance, and still more in order to get rid of a stock of unsaleable broadcloth the East India Company had on their hands.

"At first the clothing issued was so indifferent and so irregularly supplied, that the men had to supply defects themselves, thus their appearance was often tattered and always motley. The first genuine reform in the attire of the sepoys was when they were supplied with scarlet jackets of broadcloth and white linen turbans, to distinguish them from native enemies; and in 1760 the uniform of the troops in the three Presidencies was assimilated; but all had to complain bitterly of the deductions made from their pay for these necessaries.

"The earliest dress of the Madras sepoy was a coat cut away from the chest, both sides, with drawers fringed with blue, reaching to the ankle for native officers, to just below the knee for havildars, and for rank and file a few inches above the knee. A necklace of glass beads was worn round the neck, and a larger one of similar material hung down to the waist. The head-dress consisted of a blue turban in the form of a round hat, with iron plates on the four sides, and an iron rim with a rosette of linen in front. A blue cummerbund was worn round the waist." The facings of the coat were blue.

With the exception of the coat, which formed a separate charge, the articles of dress designated "slops" were kept up by stoppages from pay, sufficient to supply 3 turbans in two years, 1 sash or cummerbund per annum, 2 drawers and 2 under-jackets per annum. These stoppages did not exceed 7 fanams from a havildar, 5 fanams 12 cash from a naick, and 4 fanams from a sepoy (one rupee = 12 fanams).

"As the making up of the clothing must be left to one person's care, an agent must be appointed who must have an allowance of so much per cent. on the amount of the whole expense, equal to his trouble."

1777 In July 1777 the hats of the officers of the regiment were ordered to be "plain, cocked with gold looping; hat string and tassel as at present."

The following order, bearing the same date, shows that the regiment, although then not a year in existence, had already become a pattern in the dress of its officers to the remainder of the Madras Army:—

"Besides the full frock, the officers of sepoys are to have jackets with caps, in the manner of Captain Alcock's battalion, with a small silver plate in front, distinguishing the number of the battalion, and whether 'Carnatic' or 'Circar.' The uniform, hats, and breeches of all the officers to be the same as at present."

1780 Captain Innes-Monro gives the following description of the uniform of the Madras Infantry sepoy and native officer, in a letter written in March 1780:—

"Their uniforms have a very military appearance, consisitng of a red light infantry jacket, a white waistcoat, and a blue turban placed in a soldierlike manner upon the head, edged round with tape of the same colour with the facings, and having a tassel at the lower corner. The sepoy has a long blue sash lightly girded round his loins, the end of which, passing between his legs, is fastened behind. He wears a pair of white drawers, tightly fitted, which only come half down his thigh, and, being coloured at the lower end with a blue dye, appear as if scalloped all round; a pair of sandals upon his feet, white cross belts, a firelock and bayonet, complete the sepoy's dress.

"The dress of the black officers is much the same as described above; with this difference only, that their coats are made of scarlet cloth, with tinsel epaulettes, white drawers all the way down to their ankles, and a large crooked scimitar by their sides."

In October 1780, Government, on the recommendation of the Commander-in-Chief, directed that no stoppages should be made from the pay of the sepoys on account of "slops" or half-mounting, but that they should be permitted to provide their own.

1786 In 1786 orders were issued for the British officers to wear white covers on their caps.

1791 Captain Charles Gold, in his *Oriental Drawings* (sketched by him between the years 1791–1798), makes the following remarks about Plate II. :—

"The drawing represents a Subidar, Grenadier Sepoy, and Recruit of different regiments (Madras Native Infantry); there is no distinction in the jackets of the infantry corps, *all wearing yellow facings*, white lace, and the Sepoys without buttons. In other respects, the dress of each corps varies according to the taste of the Commandant, as is observable in their turbans, etc., preserving uniformity only in their colour." [1]

1797 In the early part of 1797 the dress of European officers was revised, and the following is an extract of the order :—

"Infantry officers are to wear black leather stocks, with a false white linen collar, one-third of an inch deep; white linen waistcoats, single breasted, and cut round as at present, with metal buttons corresponding with those on the jacket. White nankin pantaloons with half-boots, and the black round hats ornamented in such manner as officers commanding native battalions may think proper."

On the 15th March 1797 a new pattern uniform turban and blue cummerbund were ordered to be introduced for the native ranks from the 1st May; Government, in their despatch to the Directors, observing that they had given the matter "every consideration which a subject of that delicate and mportant nature required."

[1] From the above it must be inferred that the colour of the facings of the 13th Carnatic Battalion when first raised was yellow.

1801 On the 7th July 1801, Major-General Braithwaite published the following General Order:—

"The Clothing of the Honourable Company's Forces under the Presidency of Fort St. George, is established as follows:—

. . . .

"3RD NATIVE REGIMENT.

Colour of lappels, cuffs, and collar	Red.
Colour of lace	White, with a red stripe.
Colour of clothing of drums and fifes, including lining	White with red.
Colour of officers' buttons, and how set on	White, equal distances.
Officers' lace	Silver.

. . . .

"Clothing of regiments of infantry to be *jackets*—lining white, with a Grenade or Bugle horn on that part of it which unites the points of the turn-up of the skirts, to distinguish the flank and light companies respectively; and a star in the same place to denote the battalion companies.

"The grenadiers to have the usual round wings of red cloth on the point of the shoulder, with six loops of the same sort of lace as on the button-holes, and a border round the bottom, with worsted fringe.

"Officers to have the number of the regiment and battalion on the *buttons*; and their jackets to be either plain or laced, at the discretion of the Colonel.

"The *jackets* to be made full so as to button down to the waist, and to be well sloped off from the waist across the hips to the extremity of the skirt where the points meet. The collar to turn down: the lappels to reach to the waist, to be three inches in breadth: the cuffs to be rounded and three and a half inches broad. The flaps of the pocket to be across the skirt, and to be sewn down, the pocket being cut in the lining. The lace to be half an inch broad.

"European and native officers' *sashes* to be of crimson silk. Those of the non-commissioned officers to be of crimson worsted, with a stripe of the colour of the facings of the regiment; with exception of such regiments as are faced with red, which are to have a white stripe. The whole to be worn round the waist over the jacket. The Company's Arms, over the number of the regiment and battalion, to be engraved on the *gorget*, which are to be either gilt or silver, according to the colour of the buttons on the uniform, and are to be worn on the breast suspended by a riband of the colour of the facings, passing round the neck under the jacket.

"*Belt-plates* to be either gilt or silver, according to the colour of the buttons on the uniform, the number of the regiment and battalion to be engraved on them; and the plates to be of such shape, and so ornamented, as the Colonel shall direct.

"*Swords* of officers to have a brass guard, pommel, and shell, gilt with gold, with the gripe or handle of silver twisted wire, and to be worn in whitened buff leather shoulder belts. Field and regimental staff officers to wear waist-belts instead of shoulder belts. The *sword-knots* of the whole to be crimson and gold in stripes.

"Field officers to wear two *epaulets*. Officers of flank companies to wear one epaulet on each shoulder; those of the grenadiers being distinguished by a Grenade, and those of the light infantry by a Bugle horn on each. Captains and subalterns of other companies to wear one epaulet on the right shoulder, with any distinction to mark the several ranks that may be established regimentally.

"European officers to wear round black *hats*, ornamented at the discretion of Colonels respectively—Black leather *stocks*, with white false collar one-third of an inch in breadth—short white *waistcoats*—white *pantaloons*—Half *Boots* and leather *gloves*—*Hair* clubbed or queued—Buttons on the hat, waistcoat, and pantaloons to be of small size, and to correspond with those on the jacket.

"Officers are permitted, on occasions not connected with duty, to wear coats and cocked hats, with breeches, shoes, and stockings. The coats to correspond in every particular part and distinction with their regimental jacket: the trimmings on the hats to be gold; the button and loop corresponding with the colour of the buttons on the uniform.

"*Sergeants and drummers* of native infantry regiments to be dressed as in the Madras European Regiment, namely, to wear round black hats, ornamented at the discretion of the Colonel; black leather stocks, with white false collars one-third of an inch in breadth; short white waistcoats; white pantaloons, and black half gaiters. Hair clubbed or queued, with a small piece of black polished leather on the club or queue.

"As the articles of half-mounting lodged in the Military Board Office for public reference do not exhibit any distinction in either for different ranks, none will be allowed; neither are any additional distinctions (beyond those prescribed in General Orders to denote the confidential men) to be made in the clothing, the several gradations of rank being sufficiently marked when delivered by the contractors.

"A plate on the *Cartouch Box* is positively forbidden as tending to injure the pouch."

Cummerbunds, as an article of uniform, were discontinued in 1803. 1803

The following General Order was published on the 29th December of the same year:—

"The Commander-in-Chief admits of a standing collar (which is to button on either side of the neck) being added to the under jacket, and that the drawers are to reach to the distance of one inch to the upper part of the knee pan."

The uniform turbans, as fixed in 1797, did not give satisfaction for long, for in 1805, on account of their weight and inconvenience, a new pattern (with leather cockade) was ordered to be taken into use.

The following General Order was published by the Commander-in-Chief on the 14th November 1805:— 1805

"Lieutenant-General Sir J. F. Cradock has established a turban for the native commissioned officers, non-commissioned officers, and rank and file of regiments of native infantry. Sealed patterns will be furnished. Each turban is to be made sufficiently large to reach low down upon the head, and to fit so firmly as to prevent the turban becoming unsteady when the soldier moves at an accelerated pace. The turban shall be worn even upon the head, and touching the eyebrows."

The men, however, declared that the new pattern bore an offensive resemblance to the cap worn by the East Indian drummers; and, thinking that their introduction was the thin end of the wedge towards compelling them to embrace Christianity, they positively refused to wear them. The Commander-in-Chief insisted, and this led to the mutiny at Vellore.[1] Eventually the order was cancelled, and the 1797 pattern turban remained in use (G.O., dated 17th July 1806).

1806 On the 24th September 1806 the wearing of leather cockades and plumes was prohibited, and stocks of every description were abolished. At the same time the native ranks were permitted to wear their marks of caste at all times, and in any manner they thought proper. Permission was also accorded to them with respect to wearing the hair on the upper lip, the wearing of joys, and ornaments peculiar to different families and castes, which had been prohibited in the "New Regulations of 1806."

[1] "The offensive article of the regulations which occasioned so much mischief, and which has since been rescinded, ran in the following words:—'10th.—It is ordered by the regulations that a native soldier shall not mark his face to denote his caste, or wear ear-rings when dressed in his uniform. And it is further directed that at all parades, and upon all duties, every soldier of the battalion shall be clean shaved on the chin. It is directed also that uniformity, as far as practicable, be preserved in regard to the quantity and shape of the hair upon the upper lip.' This trifling regulation, and a turban with something in its shape or decorations to which the sepoys are extremely averse, were thought to be so essential to the stability of our power in this country, that it was resolved to introduce them, at the hazard of throwing our native army into rebellion. A stranger who reads the Madras regulation would naturally suppose that the sepoys' beards descend to their girdles, and that they are bearded like the pard; but this is so far from being the case, that they are now, and have been as long as I can remember, as smooth on the chin as Europeans, making a due allowance for the difference of the razors employed on the two subjects; and as to the hair upon the upper lip, its form is so much like that which sometimes appears upon the upper lip of our own dragoons and grenadiers, that none but the critical eye of a shaver could distinguish the difference."—THOMAS MONRO. To Hindoo and Muhammedan alike the new turban, made in part of leather prepared from the skin of the unclean hog or of the sacred cow, was at once an offence and a desecration.

In 1806 a new pattern forage cap was ordered to be worn by the British officers.

The following is an amusing description of the dress of British officers about the year 1808:—

"The officers wore large cocked hats and feathers, white leather or kerseymere breeches, and long boots above the knee, like dragoons, with powder and long tails, the curl of which was generally formed of some favourite lady's hair, no matter what the colour might be. The facings of the regimental coat were buttoned back. White waistcoats, white doeskin gloves, black silk handkerchiefs (invariably tied behind), swords with white buff leather sword-belt and breastplate, were worn. Black cloth leather gaiters were used when on duty with the men, and were fastened with about twenty flat silver buttons. Gorgets were worn by officers on duty or on court-martial duty. The evening dress was grey cloth tights, with Hessian boots and tassels in front, and everyone was powdered and correctly dressed before sitting down to dinner."

About this time, the officers of the grenadier companies were ordered to wear felt hats when they did not use their proper grenadier caps. The tufts to be white.

1809 In 1809 the facings of the 1st Battalion, 3rd Regiment, were green, and of the 2nd Battalion, red. Gold lace was worn by the 1st Battalion, and silver lace by the 2nd.

1810 On the 9th March 1810 the issue of woollen cloaks and pantaloons to all native troops proceeding on foreign service was authorised.

The following are extracts from an Order published by the Commander-in-Chief in October 1810:—

"Clubs and queues are abolished in all ranks, and the hair is in future to be cut close to the neck—no powder to be worn on duty."

"All field officers are to wear two epaulettes."

"Captains of flank companies, who have the brevet rank of field officer, are to wear wings in addition to their epaulettes."

"Captains and subalterns are to wear one epaulette on their right shoulder, excepting those belonging to flank companies. Captains and subalterns of flank companies are to wear a wing on each shoulder, with a grenade or bugle horn on the strap, according as they belong to the grenadiers or light company."

"In lieu of pantaloons and half-boots, wide trousers with gaiters are permitted to be worn, but all field officers will still wear full boots and chain spurs."

1811 On the 30th September 1811 a pattern knapsack

and havresack was introduced; the latter being intended chiefly for the carriage of rice, for which purpose the former was forbidden to be used.

The knapsack was to contain "3 white jackets, 3 pairs of drawers, 2 pairs of knee bands, 2 handkerchiefs, 2 duppatahs, 2 loongies, 1 flat dish, and 2 basins (of a portable size fitting into each other), emery, whiting, blacking or heelball, and pipeclay, also one small carpet neatly folded and placed on top of the knapsack, over which the watch-coat was to lie horizontally, both to be attached to the knapsack by the long straps. The jumboo was to be slung from the sepoy's right shoulder."

1812 As the short drawers in use with the native ranks were found to be inconvenient, although they had been worn for nearly half a century, it was decided on the 22nd January 1812 to permit pantaloons to be worn instead.

The following is an extract from a letter, dated 4th February 1812, on the subject:—

"The white pantaloons are to be considered as the full dress of the battalion, and to be worn on all occasions of parade, in review order, general duties of garrison and cantonments, and on occasions of ceremony. Each man to be provided with two pairs. The coloured pantaloons to be made of striped soussee, or such other suitable cloth as the Commanding Officer may, from the local situation of the corps, think it eligible to adopt, but the greatest attention must be paid to uniformity in colour. Commanding Officers will take care to choose the most plain, and correspondent either with the colour of the jackets or facings."

The same letter contained the following orders about sandals:—

"The Commander-in-Chief also directs that you will take every pains to establish a uniform sandal to be worn by the men on all duties, and that it may be fixed upon, in communication with the native officers, of a light strong muster, sufficiently large in the sole to cover the bottom of the foot, and not encumbered by any unnecessary straps or fastenings over the foot."

1814 In July 1814 the British officers were ordered to wear "Wellington pantaloons and half-boots and black kidskin caps" (being, however, unsuited to the climate, they were abolished in 1816). Grey cloth pantaloons could be worn in bad weather. Sword-belts and sashes were ordered to be worn above the greatcoats. Officers were directed to wear undress jackets and foraging caps only at drills and on duty within their respective lines. The appearance of any officer in public except in full dress was prohibited. 1816

In June 1816 a new regulation cap for the British officers was introduced. This was, however, superseded by a beaver cap in February 1822.

During September uniform shell jackets were ordered to be worn at drills, fatigue duties, marching, and in the lines.

The following is an extract from a General Order published by the Commander-in-Chief, dated Fort St. George, 3rd February 1817:— 1817

"The loose overalls (the present established uniform of officers) are not considered as appointments fitting for occasions of ceremony, for a ball-room, or evening dress; but white pantaloons, and half-boots over them, may be worn on such occasions by all officers.

"When officers in evening full dress wear shoes, they are to wear shoe buckles and white breeches, which should be established regimentally—strings in the shoes or at the knees are prohibited, and it must be understood that, in the full dress, the sash is never to be worn.

"Cocked hats and long coats, according to regulation, are only permitted to be worn in evening dress with shoes and stockings or pantaloons and half-boots, as above described.

"The foraging cap and undress jacket are to be worn only on occasions quite unconnected with duty or ceremony, and it is to be understood that officers are not to appear abroad, in public places, at the Presidency or other stations, except in the full-established regimentals of their respective corps."

On the 17th February 1818 a Clothing Board was established at Fort St. George, for the purpose of conducting all business connected with the clothing and half-mounting of the army. 1818

On the 17th February 1819 the issue of woollen trousers to the native ranks was sanctioned, at the rate of one jacket or one pair of trousers in each year. This issue was granted owing to "the sufferings endured by the native troops from want of sufficient clothing, more 1819 especially when serving beyond the frontier, where they were exposed to great changes of temperature." These articles were to be paid for out of the Off-Reckoning Fund.[1]

1823 On the 12th February 1823 three orders of dress were introduced for British officers of Native Infantry Regiments, namely, "full dress," "dress," and "undress."

In "full dress" a scarlet coatee, with Prussian collar three inches deep with a small button at each end, buttoning back to ten regimental silver buttons, was worn. The coatee skirts had cross flaps and white kerseymere turnbacks with regimental skirt ornament. The 2nd Battalion, 3rd Regiment, was authorised to wear lace on the turnbacks and the back skirt. Gold epaulets were worn on the coatee, the officers of the flank companies wearing wings. The cap consisted of a black bell-shaped beaver, with felt crown seven and a half inches deep, and lacquered top with a star in front bearing the device of the regiment, surmounted by a cockade. A feather with red and white upright hackle, twelve inches long, with a gilt socket, was attached to the cap. The breeches were made of white kerseymere, with regimental buttons and buckles at the knee. White silk stockings, black silk cravat, and shoes with plated buckles, were worn. The sword-scabbard was made of black leather.

In "undress," a scarlet round shell jacket was worn, which was quite plain, with the exception of a scarlet silk twist on each shoulder. White linen trousers and an oilskin covered cap were worn in this order.

The "dress" was the same as full dress, except that the lappels were buttoned over, and loose white linen trousers and Wellington boots were worn. A white buffalo leather shoulder-belt, three inches wide, was

[1] See Part IX. MISCELLANEOUS, for an account of this Fund.

worn diagonally across the shoulder. A crimson net silk sash was also worn, which was folded twice round the body and tied at the side.

Mounted officers were ordered to provide themselves with a saddle-cloth of the same colour as the regimental facings, 2 ft. 2 in. long, 1 ft. 7 in. deep, with gold lace $\frac{5}{8}$ in. wide, edged with the same colour as the coat, a black leather bridle, a white collar, and holster covered with black patent leather.

A General Order, dated 19th June 1824, directed that, "with reference to the recent alteration in the numbering of corps of infantry, the several regiments of infantry, according to their new numbers, will retain the colour of facings and lace, each respectively wore as a battalion of a regiment."

1825 In 1825 the native ranks wore "red coats, white trousers, blue turbans, with a brass ball at the top."

1827 The following Order was published by the Commander-in-Chief on the 20th November 1827:—

> "White jackets and white cravats are forbidden to be worn by an officer at any time out of quarters. Silk or crape jackets or trousers are also prohibited; broadcloth being the established material of which officers' clothing is to be made; all deviations from established regulations, whether in quality or fashion, are prohibited. Should officers disregard these orders, and appear again in white cotton jackets, or fancy clothing of any kind, the Commander-in-Chief will prohibit the shell jacket and forage cap being worn, and order them to appear at all times in the uniform established for the parade."

1828 On the 26th March 1828 orders were issued that officers and men of all dismounted corps, whether artillery, light infantry, pioneers, or infantry, were to wear trousers of the colour called "infantry blue-grey."

1830 The Commander-in-Chief issued orders on the 25th June 1830, that neither mustaches nor the beard on the chin were to be worn, except by the native ranks.

1832 On the 6th December 1832 it was ordered that the shoulder-belt with slings should be no longer worn by field officers, and that they should wear a buffalo leather sling waist-belt two inches wide, with regimental plate, and a brass instead of a leather or steel scabbard. They were, however, directed to retain the black waist-belt for undress.

1839 The facings of the regiment were changed from red to white by G.O.C.C., dated 2nd December 1839.

1840 On the 9th March 1840 the following Order was promulgated :—

> "As there is a great want of uniformity in the manner in which mourning is worn by officers, the Officer Commanding-in-Chief is pleased to direct that, except at funerals, no other shall be used than black crape three inches broad above the elbow of the left arm, and the sword knot also covered with black crape. At funerals, officers may wear black crape scarfs over the shoulder, and bands on the helmet or chakoe."

In 1840 the musicians and drummers wore white tunics with red facings, gold lace and epaulets, dark blue trousers with broad red stripes down the sides, large red chakoes with a white plume, and black shoes with brass buckles. The colour of the tunic was changed from white to red about ten years afterwards.

1842 On the 2nd February 1842, surgeons and assistant-surgeons were directed to wear the uniform of the regiment to which they stood appointed. Epaulets to be of gold as regulated for captains and subalterns respectively. Forage cap to have the regimental badge or number. The lace to be worn by medical subordinates was also to be gold instead of silver, and the buttons to be gilt instead of plated.

1851 Packs instead of knapsacks were taken into use by the regiment in 1851.

1852 In 1852 British officers wore the coatee, white shell jacket, and chaco, the coatee in full dress, the shell jacket in drill order.

By an Order dated 25th July of this year, the British officers were permitted to wear mustaches.

The following is an amusing description of the dress of the native ranks prior to 1853:— 1853

"Knapsacks, basket hats, stocks, tight clothing, pipe-clayed belts, and other unspeakable absurdities, introduced in imitation of European follies, are quite sufficient to crush and cripple the soldiers, without much effort on the part of the enemy. The assimilation of everything to a European model, however absurd that model may be, has enabled the native army to accord in all outward appearances, ceremonies, and forms, to the European armies, but its real efficiency has thereby been destroyed.

"A sepoy of the line dressed in a tight coat; trousers in which he can scarcely walk and cannot stoop at all; bound to an immense and totally useless knapsack, so that he can scarcely breathe; strapped, belted, and pipe-clayed within a hair's-breadth of his life; with a rigid basket chako on his head, which requires the skill of a juggler to balance there, and which cuts deep into his brow if worn for an hour; and with a leather stock round his neck, to complete his absurd costume—when compared with the same sepoy, clothed, armed, and accoutred solely with regard to his comfort and efficiency, forms the most perfect example of what is madly called the 'regular' system with many European officers, contrasted with the system of common sense now recommended for adoption."

On the 10th October 1853 the leather stock, which had still been retained for full dress, was ordered "to be discontinued, with the cloth jackets or coatees at all times, and with the white jackets when the men were on duty; nothing was to be worn inside the coat beyond a plain white shirt collar of sufficient depth to prevent the soiling of the lining of the coat collar, without being visible above it."

The men, when not on duty, were permitted to wear the white jacket open (with the exception of the lower buttons), with white shirt and a black silk neckerchief.

When on service the collar of the coat or jacket was to be unhooked, the upper buttons opened, and the collar turned back, so as to give the sepoy as much freedom and relief as possible.

"When the power of the sun rendered more than usual protection to the head and back necessary, the European officers and the buglers, drummers, and fifers were permitted to wear a helmet forage cap, composed of pith or some other light and non-conducting material with a white cover, to be wadded throughout with cotton, with a flap or curtain to protect the sides and back of the head. When forage caps of the above description could not be readily procured, the best substitute attainable, such as thick bands of white cloth rolled round the chako or forage cap, should be adopted, so as to secure the desired object—namely, the preservation of health and life—appearances being disregarded, so far as it may be necessary so to do. The native officers and soldiers were also permitted to use similar cloths round their turbans."

1854 On the 18th November 1854, "helmet forage caps" were sanctioned to be worn by British officers on parade, when the white turban covers were worn by the men.

1856 The adoption of tunics, waist-belts, and sashes similar to those prescribed for officers of Her Majesty's service, was authorised in October 1856 for the British officers of the regiment.

1859 The Commander-in-Chief sanctioned the adoption, by all British officers of infantry, of the khaki costume as their summer clothing, on the 17th August 1859. The material was not restricted to cotton, but cashmere, linen, or other suitable material could be adopted, provided it was of the proper colour, and that uniformity was observed regimentally. The khaki costume could be worn on all duties on which the white had been worn. The forage cap or wicker helmet could be worn with the khaki clothing, the cover being of khaki colour. The above order did not apply to the native ranks.

The light felt or wicker helmet was ordered to be worn by British officers, with a white quilted cover and a turban of khaki or unbleached muslin, after August 1859.

After the abolition of stocks for full dress, a small black neckcloth of a uniform pattern was ordered to be worn. With the summer clothing no neckcloth was to be worn.

1860 "On the 3rd March 1860, the introduction of a new head-dress for the native ranks, in lieu of that previously worn, was sanctioned by the Home Government. It was to consist of a plain native turban of muslin, or of cloth, the pattern and colour to be established regimentally. It was left to Commanding Officers to introduce it or not, according to their discretion."

Tunics were first issued to the regiment on the 8th March 1860. The brass buttons on these cloth tunics were embossed with the numeral 13 within a laurel wreath.

After sanction had been accorded to the introduction of a new turban in 1860, a turban of black varnished cloth was chosen for the regimental pattern. It was worn with the white jacket, which was fastened by six lead buttons embossed with the numeral 13 in a laurel wreath. Black trousers and brown leather sandals were also worn in this dress, which was usual for ordinary parades and guards.

For parades in review order the black turban was covered with white cloth, and a black braided band with a tassel at the right side was hooked upon it in front. A red tunic, with six blue and white tapes in front, fastened with six large brass buttons embossed with the numeral 13 in a laurel wreath, was also worn in this order. The red tunics of the flank companies had white wings on the shoulders. White trousers and brown leather sandals were worn by all native ranks in review order.

On fatigues and roll-call parades the men wore a white cap on the right side of the head, with two red bands running round it. The cap was of the same shape as the present officers' active service cap, namely, the Austrian pattern.

1861

In 1861, when the regiment was quartered in Trichinopoly, the colour of the regimental turban was changed to red. The new red turban was worn in review order only.

The white jacket, or "angreka," was worn on all other duties and parades, and was ordered to be made to fit with freedom, but not too loosely, particularly the sleeves, which were not to be larger than necessary at the wrist. Black linen trousers were worn with the "angreka." These were ordered to be made to fit neatly about the hips without being too tight, and to be made straight down from the hip to the instep, perfectly loose, so as to give the men a "full appearance."

It was particularly to be observed that the trouser reached freely to the instep, without, however, being too long or bagging, and that it was to be of the proper breadth at the bottom, as the men were apt to narrow it towards the ankle. The white cap, with the two red bands described above, and leather sandals, completed this dress.

1863 In G.O.C.C., No. 101, dated 18th November 1863, "His Excellency the Commander-in-Chief desires that the red sash be invariably worn by officers (of infantry) when dressed in white clothing on all occasions of duty and on parade."

1864 In Cannanore in 1864 the colour of the ordinary duty and parade cap was changed to red, with a white band running round it. No change was made in the white "angreka" and parade trousers.

1865 On the 26th January 1865, British officers of native infantry regiments were ordered to wear a red serge frock,—to have the facings of the regiment,—badges according to rank to be worn on the collars. Field officers were to be distinguished by a row of gold braid at the top and bottom. Company officers were to wear the braid at the top only; the red sash always to be worn with it. The frock to be worn at such times as General Officers directed.

1866 On the 8th August 1866 the following articles of kit were ordered to be kept up by native officers, non-commissioned officers, and men:—1 turban, 1 cloak (of cloth or blanket), 1 valise, 1 carpet, 3 white jackets, 3 pairs white trousers, 2 pairs black trousers, sandals, shoes, or half-boots of European pattern, brass dishes, and 1 havresack.

The recruit and pension boys' kit was to consist of a skull cap of white or red turkey twill, white jackets similar to those of the men, and without collars, blue or black trousers, and sandals. They were to wear a sash of black wove linen cloth, with tassels tied on the right side, and with the tassels hanging down the front of the right hip. The sash was to be tied on the natural waist, just above the hips, so as not to press upon the ribs, and was not to be tied broader than its own width.

The native ranks were also to have 3 "angrekas," 1 loonga, 1 dooputtah, and 1 kerchief or skull cap, which were to be kept up out of their pay.

The articles were carried in a bag of regimental pattern, with the exception of one suit of clean clothes, which were carried in the waterproof cover.

The drummers' and musicians' kit included five cotton shirts, a glengarry, and a brush and comb.

On the 15th November 1866 orders were issued for a button to be added to the havresacks on the side next the wearer, an inch lower down than the button, which fastened the flap. When empty, the havresack was to be worn rolled up and slung across the right shoulder under the bayonet.

1867

In March 1867 the blue double-breasted frock coat (with which white trousers were worn) was abolished for British officers, and a blue patrol jacket, with mohair braid, double drop loops, and a row of knotted olivets, substituted. The sword-belt was ordered to be worn under the new patrol jacket.

1868

On the 9th May 1868 officers of the Staff Corps were ordered to provide themselves with a single-breasted scarlet tunic, with blue collar and cuffs, and with a crimson silk cord on the left shoulder to retain the sash. The tunic to be fastened in front by eight buttons equidistant. The buttons to be embossed with the cypher V.R. in a garter surmounted by a crown, with the words, "Madras Staff Corps" in a garter. The collar badges to be a crown and star for a colonel, a crown for a lieutenant-colonel, a star for a major, a crown and star (with unbraided eyes) for a captain, and a crown (with unbraided eyes) for a lieutenant. The *hat* was to be "cocked," the fan or back part 9 inches deep, the front 7½ inches, each corner 5 inches, gold lace loop, and tassels of gold crape fringe, with crimson crape fringer underneath. *Feather*, white and upright; hackle 5 inches long. *Stock*, black silk, or a black silk neckerchief. Gold lace accoutrements and trouser stripes were ordered to be worn this year by all officers of infantry when in full dress.

On the 16th June 1868 the native ranks were ordered to wear a collar made of three plies of white cloth, to fit the neck, having two rows of stitching all round, and buttoning behind with three buttons, one-eighth of an inch of white collar to appear above the collar of cloth tunic. The wearing of these collars was discontinued in this regiment in 1875.

1869 With the sanction of Government, the Commander-in-Chief was pleased to abolish white trousers as an article of clothing for the native ranks of the army on 5th January 1869.

1870 On the 5th August 1870 a patent cork helmet, covered with white cloth, as manufactured by Messrs. Hawkes & Co., was ordered to be worn by European officers of native regiments. The helmet was to be without ornament of any sort (with puggree of regimental pattern), and with a curb chin strap gilt for officers dressed in red.

When the regiment was in Hong Kong in 1870, the native ranks wore a round red cap, with a white band at the bottom and a white bob on the top. The cap was worn over the forehead, and not on one side. In this year the wearing of leather sandals was discontinued, and black leather European boots taken into use, by all ranks of the regiment.

1871 On the 6th December 1871 waterproof kit bags were introduced into the native army.

When the regiment was stationed at Madras in 1871, the native ranks wore the white coat without collars, black trousers, and the round red cap with white band described above, on all duties and parades, with the exception of parades in review order, when the red turban, red jacket, and black trousers were worn.

1872

On the 3rd June 1872, officers were instructed to provide themselves with a plain white tunic, in lieu of the braided one, to be of sufficient length to just clear the saddle when mounted, to be fastened by four small gilt buttons of regimental pattern, cuffs pointed, and pocket outside left breast, collar rounded off in front, twisted gold cord on each shoulder. The sash to be worn on all occasions except at mess. Field officers were to wear distinctive badges on collar.

On the 15th November 1872, the clothing of bands of native corps was changed from red to white, commencing from issue for 1874–75.

1876

In 1876 the native officers wore the red tunic, red turban, sash, cloth trousers, gloves, and sword-belts in review order; other ranks, the red tunic and head-dress and cloth trousers.

Marching order was the same as review order, except that the men wore a white havresack, cloak, and complete kit in a waterproof cover.

In drill and field order all ranks wore the same articles as in review order, but with cloth or linen trousers, according to season.

The white jacket or "angreka" could, however, be worn on all parades and duties, if sanctioned by the local Commanding Officer.

When detailed for fatigue parties, the men wore the "angreka," cloth or linen trousers according to season, and the skull cap, if regimentally introduced and sanctioned by Headquarters. Native officers with such parties were dressed in field-day order.

The collar was only worn with the tunic, and consisted of three plies of white cloth as introduced in 1868.

The waterproof cover was a plain piece of waterproof cloth, two and a half feet square, lined with white cloth.

The cloak was folded flat, and strapped on the man's back outside the waterproof cover, which contained certain articles and the carpet.

The jumboo was fastened on the cloak or kept in the havresack.

The havresack was made of strong white cloth, slung over the right shoulder and above all the accoutrements, hanging down the left side two inches below the hip-joint.

To ensure uniformity and adherence to orders in the various particulars of dress, there were in each company two pattern men. Wing commandants selected them, and were held responsible that every man was dressed and equipped in exact conformity with the pattern thus established. Pattern men were paraded monthly for the Commandant's inspection, and were afterwards sent to their companies for inspection by the men. This custom was discontinued in 1887.

All branches of the native army were supplied with clothing made up, with the exception of that for native officers and for one havildar and five rank and file per company, which could be issued in materials, or if so desired, in bulk, uncut and untrimmed.

On the 20th March 1876, the issue of serge trousers in substitution of cloth trousers was sanctioned.

On the 23rd June the wearing of white cap covers by Native Infantry Regiments was ordered to be discontinued, as "they were unnecessary, and when somewhat dirty or put on carelessly, they gave a slovenly air to the men, and detracted from that soldierlike appearance which Commanding Officers should try to establish and maintain in their regiments."

On the 14th October serge frocks were ordered to be substituted for cloth tunics, with the exception of one cloth tunic issued every four years. An extra serge frock was sanctioned from the saving accruing from the change. Thus the clothing for native infantry was one cloth tunic, two serge frocks, and two pairs of serge trousers every four years.

1877

On the 29th January 1877 the pattern of waterproof kit bag, sanctioned on the 6th December 1871, was abolished, and in lieu thereof, stout canvas bags were provided regimentally.

In 1877 a recruit received thirty rupees kit money on joining, and the sepoy four rupees every year for half-mounting.

Serge trousers were issued to the regiment on the 1st April 1877, at Madras.

1878

Serge trousers and serge frocks were issued to the regiment on the 1st April 1878, and orders were issued for the serge frocks to be worn in marching, field day, and drill order. The brass buttons on the serge frocks were inscribed with the royal crown and the cypher V.R. The buttons which had been received with the tunics in 1860 were now worn with the white jackets.

By S.G.O., No. 31, of 1878, medical officers, when performing mounted duties, were ordered to wear knee boots and pantaloons.

By S.G.O., No. 77, of 1878, the departmental turban for the Indian Medical Service was ordered to be white, but officers of the department when attached to regiments were ordered to wear the head-dress of the corps with which they were serving.

By S.G.O., No. 99, of 1878, the introduction of cork helmets as the head-dress of drummers, buglers, and musicians of native corps was authorised.

By S.G.O., No. 106, of 1878, regimental badges were sanctioned to be worn by officers on the helmets. "The 'front plate' of the helmet was not to be worn in place of the puggree."

1879 When the regiment was stationed in Meean Meer in 1879, the red turban was discarded, and a black turban with khaki band was adopted in its place.

On the 24th April 1879 the armourer havildar was permitted to wear the distinguishing chevrons of his rank.

1880 By S.G.O., No. 5, dated 20th January 1880, native officers were ordered to wear a puggree of regimental pattern.

1881 On the 28th February 1881, British officers were ordered to wear shoulder straps of universal pattern twisted round gold cord, and lined with scarlet, on their tunics and shell jackets. The shoulder straps of patrol jackets, greatcoats, etc., were to be of the same material as the garment.

Badges of rank in silver for the tunics and shell jackets were ordered to be worn on the shoulder straps. Badges of rank in gold on the shoulder straps of the patrol jacket and greatcoat.

The badges of rank for a colonel were a crown and two stars, for a lieutenant-colonel a crown and one star, for a major a crown, for a captain two stars, for a lieutenant one star, and no badge for a second lieutenant—as at present.

On the 30th April 1881 orders were issued for the extra chevrons and badges to be removed from the left arms of tunics, frocks, jackets, etc., and the good conduct badges removed from the right arms to the left, and all four bar chevrons to be worn below the elbow.

The chevrons below the elbow were all reversed, that is, the points now to be upwards instead of downwards, and the chevrons, etc., for the drum-major to be removed from above to below the elbow.

By S.G.O., No. 2, of 1881, a new pattern forage cap for British officers was sanctioned. It was made of blue cloth, straight up, three inches high, with black patent leather drooping peak, and chin strap. The peak was ornamented with half-inch full gold embroidery; black netted button and braided figure on crown, etc.

At the same time a cap for active service and peace manœuvres was introduced. It consisted of a cap of blue glengarry pattern, bound one inch wide with black silk riband. A black silk cockade was on the left side. Numbers or badges were worn on the cockade with a scarlet edging.

On the 30th May 1881 gold badges of rank were ordered to be worn on the shoulder straps by all officers in summer uniform.

New waterproof helmets were sanctioned to drummers, buglers, and musicians on the 28th November 1881.

1882

On the 14th September 1882, the facings of the native regiments were limited to white, green, and yellow. The 13th Madras Native Infantry were permitted to continue wearing white facings; the 20th Madras Native Infantry, white; and the 22nd Madras Native Infantry, yellow. The facings of the latter were not assimilated to the former when all three were linked together in October 1886.

When the regiment was in Jubbulpur in 1882, the khaki band, which had been worn with the turban, was removed.

The white tunics of the native ranks were ordered to be dyed khaki in 1882. This, however, was not a success, as the dye was not lasting, and after a few washings there was little uniformity of colour.

1883

"On the 17th May 1883, the old dress of the native

army was given up, and the following articles substituted :—One serge zouave jacket, one serge knickerbockers, one pair of khaki gaiters, one khaki blouse, one khaki knickerbockers and gaiters, one khaki turban, with band of colour of regimental facings, and a fringe of the same colour as the turban."

The turbans were ordered to be tied uniformly, according to the regimental pattern, and were not to be stitched or pinned.

The khaki blouse was to be made to fit loosely, in order to give perfect freedom of movement, and to permit of the serge jacket, or other warm clothing, being worn underneath. The khaki knickerbockers were to be made to fit loosely, in order to permit of a man stooping, sitting, or working in any position, as also to permit of warm clothing being worn with them. When worn without putties or gaiters, the bottom of each leg of the knickerbocker was to reach the ankle.

Khaki clothing was first taken into wear by all ranks of the regiment at Jubbulpur in March 1883, and orders were received that it should be worn instead of the white summer uniform, and on such other occasions as may be directed by General Officers Commanding Districts. Khaki clothing had been introduced into the European regiments of the Honourable Company's army so far back as 26th June 1858.

On the introduction of the khaki blouse, instructions were issued for the buttons to be covered with khaki; but in this and in some other regiments the old regimental buttons bearing the number of the regiment continued to be worn, being superseded eventually by those of the universal pattern, embossed with the royal crown and cypher V.R.

On the 17th May 1883 an order was published to the effect that the turban fringes of native officers could be of gold, if preferred. It was not, however, until seven years afterwards that gold fringes were adopted by the native officers of this regiment. At this time Government sanctioned a triennial issue of a serge zouave jacket, a pair of serge knickerbockers, and khaki material for a pair of gaiters.

In addition to the annual grant of four rupees per man, a free issue of khaki material to the value of three rupees per man, for all ranks excepting native officers, was ordered to be issued annually on the 1st April by the Clothing Department—this material to be applied to the provision of a khaki blouse, a pair of khaki knickerbockers, and a pair of khaki gaiters.

On the 13th December 1883 drummers and musicians were ordered to wear the turban prescribed for their regiment, in lieu of the helmet which had been previously worn by them. They were ordered to wear the turban on all parades and duties with other men of the regiment, the glengarry cap being worn on all other occasions. In lieu of the previous free issue of helmets, orders were issued that they should receive a biennial turban allowance of three rupees per man.

The native officers of the regiment wore the serge frock and knickerbockers, blue turban, sash, and khaki gaiters in review order. The havildars wore the same articles as the men, with the exception of the sash.

In marching order, the native officers wore a khaki jacket and knickerbockers, khaki gaiters (afterwards changed to putties), and a blue turban. Other ranks wore the khaki blouse and knickerbockers, putties, turban, havresack, cloak, and change of clothing in the waterproof cover.

In 1883 the officers of the regiment were ordered to carry whistles in uniform.

1884

The 13th Madras Native Infantry was ordered to wear a khaki coloured puggree, with white band four inches wide and khaki fringe, by G.O.C.C., No. 141, dated 20th March 1884. It was not, however, until six years later that the blue puggree was discarded, and this one taken into use, owing to the regiment being employed on field service.

In November 1884 the necessaries to be maintained by native infantry were fixed at 1 blanket, 2 pairs of boots, 3 khaki suits, 1 carpet, 1 greatcoat (of cloth or blanket), 1 khaki havresack, 1 pair putties, 1 turban, 1 water-bottle. The musicians were to be provided with 3 cotton shirts, 1 glengarry, 1 comb and brush, and 1 canteen. The drummers and musicians obtained glengarry caps in Jubbulpur in 1884.

1885 The water-bottle became a regimental necessary for the first time in 1885. It was covered with khaki drill, and slung over the left shoulder by a khaki tape.

1886 On the 31st May 1886 the non-commissioned officers of the native army were ordered to wear badges corresponding to each rate of good conduct pay drawn under the Pay Code. The badges were supplied by the Army Clothing Department, or they could be provided, under regimental arrangements, at the expense of the State.

In 1886 shoulder straps were ordered to be added to the khaki blouse.

On the 16th July 1886 it was ordered that brass numerals, showing the number of the regiment, should be worn on the shoulder straps by non-commissioned officers and men, in khaki as well as in cloth uniform.

1887 From the 1st January 1887 the half-mounting allowance was increased from Rs.4 to Rs.5 per annum, to be drawn quarterly in arrears.

On the 27th May 1887, Government sanctioned Rs.2, 8a. per man proceeding on active service, in lieu of a pair of boots.

On the 12th August 1887 a new pattern khaki coat was sanctioned for officers.

On the 31st October 1887 the following additional

regimental necessaries were to be kept up to govern claim for compensation :—2 pairs of socks, 1 jumboo and strap, 1 set of dishes, 1 waterproof pack-cloth, 1 piece of rope for jumboo, 2 pairs khaki gaiters, and 1 brush.

1888 On the 28th September 1888 the adoption by the native officers of shoulder straps and badges of rank, in substitution of the red silk worsted cords and badges previously worn, was sanctioned :—

Subadar-major .	An embroidered silver crown on each shoulder strap.
Subadar .	. Two embroidered silver stars on each shoulder strap.
Jemadar .	. One embroidered silver star on each shoulder strap.

These badges were granted free as a first issue.

Previous to this the subadar-major wore a gold crown, the subadars two gold crossed swords, and the jemadars gold single swords, at the top of the breast, the nearest part of the badges being quarter of an inch clear of the gold lace, three-quarters of an inch round the neck and down the front of the breast.

1889 On the 30th April 1889 a new pattern khaki (drill or serge) patrol-shaped jacket was sanctioned. The British officer's helmet was to have a zinc button covered with khaki to be used instead of the spike. Mounted officers were to wear leather chin straps. The Sam Browne belt to be the usual parade belt when khaki was worn; but in review order, and on dismounted duties in cantonment, or when the sash was worn, the sword-belt with slings was to be worn under the coat.

Buglers, drummers, fifers, and musicians were allowed to adopt canteens, or set of dishes, as preferred, on the 14th June 1889. The pattern to be uniform. Canteens were adopted by the 13th Madras Infantry.

1890 On the 28th March 1890 an annual compensation of three rupees, in lieu of khaki material, was sanctioned for all ranks, except native officers, beginning with April 1890. The money was ordered to be expended by Commanding Officers in providing a serviceable fast-dyed

15

material. In no case could cash payment be made to the soldier.

In 1890 all ranks in the regiment took khaki-coloured woollen putties into wear, in lieu of the khaki thread putties.

In 1890 the sixteen non-commissioned officers and men who had passed as regimental signallers, were each awarded a badge of crossed flags worked in worsted, as sanctioned in Clause 112 of I.A.C. of July 1889. The badges were obtained from the Superintendent of Army Clothing, Madras, and were issued annually to the best sixteen signallers who had maintained the standard of efficiency.

The retention of the regimental uniform for native officers on retirement was sanctioned on the 7th November 1890; and of the sword-belt and knot and sword (if the property of Government) after the 23rd January of the following year.

1891 On the 10th April 1891 instructions regarding the "Orders of Dress" for non-commissioned officers and men of native infantry regiments were issued. The tunic, gaiters, knickerbockers, medals (sash for native officers and non-commissioned officers), and puggree were to be worn in review order, and khaki coat and knickerbockers in review order in the hot weather. The khaki drill coat and knickerbockers, puggree, and putties, with complete kit and accoutrements, including greatcoat, havresack, and water-bottle, were to be worn in marching and field service order, at all times.

On the 18th December 1891, Government sanctioned the discontinuance of the issue of khaki material for a pair of gaiters, previously issued triennially to Madras Infantry Regiments; and directed that the annual value thereof, namely, one anna and four pies, be added to the annual compensation of Rs.3 in lieu of khaki material. This order to have effect for 1892–93.

Regiments of Madras Infantry were ordered to wear brown boots from the 25th March 1892. 1892

A free issue of whistles to all havildars was sanctioned on the 29th April 1892.

In 1893 the recruit and pension boys of the regiment were supplied with turbans (a sepoy's turban being cut in two), in lieu of the round red caps with the white band, which they had previously worn, and which their fathers before them had worn in Hong Kong. 1893

Waterproof pack-cloths for the native army were abolished on the 30th March 1894. 1894

A new pattern of khaki blouse was ordered in S.G.O., No. 26, of 1894, page 153, the principal alterations being that the buttons were changed from khaki covered ones to brass, and khaki putties for the boys being substituted for khaki gaiters.

His Excellency the Commander-in-Chief having approved of an improved water-bottle on the 2nd May 1894, called "The Burma No. 2," made of galvanised iron, this pattern was taken into use in the regiment, in supercession of the copper ones which had been supplied on payment by the arsenal. The new bottle was worn without a cover.

On the 22nd March 1895 wing officers of Native Infantry Regiments, when mounted on non-ceremonial parades, were ordered to wear breeches or pantaloons with putties, ankle boots, and steel crane-necked jack spurs. When dismounted they were to wear khaki putties when parading in khaki, and blue putties when parading in red. 1895

On the 23rd July 1895 the khaki active service cap, which had been worn since 1885, was abolished for British officers, who were in future to wear the blue cloth cap of Austrian pattern, with trimmings of regimental pattern.

The steel picqueting chains for officers' chargers were

abolished in 1895, and white head-ropes taken into use instead.

1896 On the 1st January 1896 all mounted officers were ordered to wear steel spurs of swan-neck pattern, except in the evening and at levees, when brass spurs would be worn. Thus the brass spur of swan-neck pattern, till then worn by field and staff officers, was abolished.

On the 15th February 1896 a new pattern greatcoat, with detachable cape and hood, for native infantry, was introduced at the option of Commanding Officers.

In April 1896 brass distinctive badges were ordered to be worn on the white patrol jacket, an intermediate summer mess dress (cloth mess jacket, red silk cummerbund, and trousers) introduced, and the buttons on the sleeve of the khaki coat abolished.

In August 1896 the field cap for active service and peace manœuvres was ordered to be blue with gold braid for regiments dressed in red, round the top of the outer flap. The badge was to be worn on the left side.

In October 1896 it was ordered that the field service khaki coats should be made with a roll collar instead of a stand-up one.

In 1896 the pattern of the scarlet mess jacket was altered, the gold shoulder cords being replaced by cloth, the gold lace on the sleeves being discarded, and ten brass buttons placed down the front.

1897 By M.C.O., No. 269, of 7th May 1897, the "full dress clothing" of the native ranks, namely, zouave jacket and knickerbockers, at present a triennial issue, was ordered to be made a quinquennial one, the saving resulting from the change being paid annually in arrears on the 1st April (commencing 1st April 1899), at the rate of Rs.2, 9a. 7p. per native officer, Rs.1, 4a. 10p. per havildar, and 15a. 6p. per rank and file, in the shape of money compensation, to assist them in meeting the cost of their kit. At the same time, the jumboo, and strap, and rope, and the pair of khaki gaiters were ordered to be eliminated from the articles of necessaries required to be maintained by the native soldier. The table of regimental necessaries to be maintained by the soldier was also revised as follows :—

	Rs.	a.	p.
1 Turban	2	0	0
1 Greatcoat, native infantry	9	8	0
1 Detachable cape for native infantry	3	8	0
1 Hood for native infantry	1	12	0
1 Cape, British infantry pattern, for native infantry	5	0	0
1 Kit bag	2	8	0
1 Set of dishes	3	4	0
1 Carpet	1	6	0
2 Suits of khaki, *i.e.* blouse and knickerbockers (exclusive of one suit made from free issue of material), each	4	8	0
1 Pair putties (khaki)	1	0	0
1 Khaki havresack	1	0	0
2 Pairs of ankle boots (brown leather), each	6	0	0
1 Brush	0	4	0
1 Water canteen	2	0	0
1 Blanket	2	6	0
2 Pairs socks, each	0	10	0

By G.O.C.C., No. 470, of the 29th July 1897, the Government of India was pleased to approve of the new pattern mess jacket with roll collar prescribed in Paragraph 14, G.O.C.C., No. 697, of 1896, being made applicable to the British officers serving with all native infantry regiments dressed in red. The mess waistcoat was to remain as at present laid down in A.R.I., Vol. vii., omitting all gold braid, and the gilt studs ; the garment being cut plain and open, with four small gilt buttons of regimental pattern instead of hooks and eyes. The hot weather mess jacket was ordered to remain as at present.

During the same year orders were issued that the roll collar on the khaki coat was not to be made applicable to native officers.

In M.C.O., No. 523, of the 3rd September 1897, the Government of India sanctioned the lengthening of the skirt of the zouave jacket, worn by regiments of Madras

Infantry, by one and a half inches, at a cost of Rs.650, to be met from the compensation laid down in C.O., No. 269, of 1897, the rates of which are consequently revised as follows :—

		Rs.	a.	p.
Native officers	each	2	9	4
Havildars (including drum and fife major) .	,,	1	4	7
Rank and file	,,	0	15	3

PART IV

DEVICES, BADGES, AND MARKS OF DISTINCTION

IV

BADGES, DEVICES, AND OTHER MARKS OF DISTINCTION WORN BY THE REGIMENT

IN July 1777 the officers of the regiment wore caps with a small silver plate in front bearing the number 13 and the word "Carnatic," to distinguish it from the "Circar" Battalions, from which it may be inferred that Captain Alcock, the first Commandant, was one of the first to appreciate the advantages of special badges, in fostering a feeling of emulation and *esprit de corps*. 1777

On the 26th December 1820 the following General Order of the Government of Fort St. George, entitling the regiment to the honorary distinction of "SERINGAPATAM," was published:— 1820

"The Honourable the Governor in Council is pleased to permit the under-mentioned corps to bear on their appointments, and embroidered on their respective standards and colours, in the English and Persian characters, the words "Seringapatam, 4th May 1799," in commemoration of the distinguished services of those corps, or detachments of them, at the reduction of the fortress of Seringapatam on that day, viz.—

. . . .

"2nd Battalion, 3rd Regiment, Native Infantry."

. . . .

The distinction "Seringapatam" was first shown in the East India Register and Army List for 1828, on the page devoted to the 13th Native Infantry.

The following is a description of the regimental badge, which was sanctioned several years after the 2nd Battalion, 3rd Regiment, became the 13th Regiment in 1824:—

"The number 13 within a garter inscribed "Madras Infantry" with the word "Seringapatam" on a scroll underneath, the whole surrounded by a laurel wreath and surmounted by a crown, to be worn in gilt by British officers on helmet, forage cap, and active service and peace manœuvre cap. In brass by native officers on the turban."

1878 On the 27th August 1878 sanction was given to a regimental badge being worn on the helmet by British officers of all regiments. Before this time a badge had never been worn on the helmet by any regiment in India.

1880 The badges on the British officers' sabretaches were discontinued on the 20th January 1880.

1889 G.G.O., No. 250, dated Ootacamund the 3rd May 1889, republished the following General Order by the Government of India, dated 26th April 1889:—

"HONORARY DISTINCTIONS.

"No. 378.—The Most Honourable the Governor-General in Council has much satisfaction in notifying that Her Majesty the Queen, Empress, has been graciously pleased to sanction the undernoted words being inscribed on the colours and appointments of the corps named, in commemoration of their services during the campaigns in the Carnatic and Mysore in 1780–84 and 1790–92:—

"I. For services during the campaigns of 1780–84 the word

'Carnatic,'

. . . 13th Regiment of Madras Infantry. . . .

"II. For services at the battle of Sholinghur, on the 27th September 1781, the word

'Sholinghur,'

. . . 13th Regiment of Madras Infantry. . . .

"III. For services during the campaigns in Mysore in 1790–92 the word

'Mysore,'

. . . 13th Regiment of Madras Infantry. . . .

1891 G.G.O., No. 50, dated Fort St. George the 27th January 1891, republished the following General Order by the Government of India, dated 16th January 1891:—

"HONORARY DISTINCTIONS.

"No. 64.—The Governor-General in Council has much pleasure in announcing that Her Majesty the Queen, Empress of India, has been graciously pleased to permit the corps named below to bear upon their colours, standards, and appointments the words

'Burma, 1885–87,'

in commemoration of their gallant conduct during the operations resulting in the conquest of Upper Burma—

. . . .

13th Regiment of Madras Infantry."

. . . .

After permission had been accorded to the regiment to wear the above-mentioned honours on their appointments, application was made to Army Headquarters for permission to wear the following badge (in place of the one previously described), which was sanctioned by S.G.O., No. 49, of 1891, and the dress regulations altered accordingly :—

"The number 13 with 'Madras Infantry' in the centre of a star within a scroll bearing the words 'Carnatic,' 'Sholinghur,' 'Mysore,' 'Seringapatam,' and 'Burma 1885–87,' the whole surmounted by a crown, to be worn in gilt by British officers on helmet, forage cap, and active service and peace manœuvre cap—by native officers on turban. In brass on turbans of rank and file."

PART V

COLOURS

V

ALTERATIONS IN THE COLOURS OF THE REGIMENT

In July 1777, the year after the formation of this regiment, the following orders were issued with regard to colours :—"All the colours of the sepoy corps on this establishment to be uniform, distinguishing only the number of the battalion, with the word 'Carnatic' or 'Circar.'" Previous to this they had been of various colour, each company having its own colour bearing the subadar's device, such as a sabre or dagger; and those carried by the grenadier companies being distinguished from the others by such distinctive marks as crosses, parallel stripes of different colours, etc. 1777

In December 1786, at the recommendation of the Military Board, Government decided that colours should be supplied to native battalions gratis, instead of being paid for out of stoppages, as up to that time had been the case. 1786

The following is an extract from a General Order, dated Fort St. George, 5th December 1800 :— 1800

"Every new raised corps will be provided with a set of standards, or colours, at the expense of Government; which are afterwards to be kept up from the Off-Reckoning Fund, arising from effective corps of the Army.

"The provision, etc., of standards and colours is regulated as follows :—

"When a new raised corps shall have attained to a sufficient degree of perfection in its discipline to warrant the delivery to it of standards or colours, a report thereof shall be made by the Commanding Officer of the corps to Head Quarters, accompanied by an indent upon the Contractor for Army Clothing for the standards or colours required."

1801 The following General Order was published on the 7th July 1801 :—

> "1. The first colour of every regiment to be the Union Flag of the United Kingdom of Great Britain and Ireland.
>
> "2. The second colour of each regiment to be the colour of the facings of the corps, with the union in the upper canton, except those regiments faced with red, white, or black. The second colour of those regiments which are faced with red or white is to be the red cross of St. George in a white field, and the union in the upper canton. The second colour of those which are faced with black is to be St. George's cross ; union in the upper canton ; the three cantons in black.
>
> "3. In the centre of each colour is to be embroidered, in gold Roman characters, the number of the regiment within a wreath of laurel. The size of the colours to be 6 ft. 6 in. flying, and 6 ft. deep on the pike. The length of the pike (spear and ferrule included) to be 9 ft. 10 in. The cords and tassels of the whole to be crimson and gold mixed."

1820 In 1820 the regiment received colours made in conformity with the flag of the United Kingdom. The first colour, called the Queen's colour, was the union flag of Great Britain and Ireland ; the second, called the regimental colour, was the red cross of St. George in a white field, with the number 13 embroidered in the centre in golden letters within a laurel wreath. The size of the colours was 6 ft. 6 in. flying, and 6 ft. deep on the pike, the length of which, including the spear, was 9 ft. 10 in. ; the cord and tassel were crimson and gold mixed.

1829 The practice of allowing men to use the regimental colours on the occasion of their principal festivals was prohibited in January 1829 ; and, probably with the idea of avoiding invidious distinction, it was also ordered that they should not be made use of at balls or entertainments given by European officers.

1840 On the 20th November 1840 the following Orders were issued with regard to colours :—

> "The royal or first colour is to be the great union throughout, and in the centre is to be painted or embroidered in gold the number of the regiment, surmounted by a crown, together with any motto which it may be authorised to add thereto, within the wreath of roses, thistles, and shamrocks, on the same stalk.

"The regimental or second colour to be of the colour of the facings of the regiment, with the union in the upper canton, except where facings are red, white, or black. If faced with red or white, the colour is to be the red cross of St. George in a white field, and the union in the upper inner canton; if faced with black, it is to be St. George's cross throughout, the union in the upper canton, the three other cantons black. The second colour is to bear the number of the regiment, honorary distinctions being borne on the regimental colour only.

"Size of the colour to be 5 ft. 10 in. flying, and 4 ft. 6 in. deep on the pike, the length of the pike, spear and ferrule included, to be 9 ft. 6 in.; the cords and tassels of the whole to be crimson and gold mixed."

1842 New colours were presented to the regiment, on the 12th December 1842, by Major-General G. M. Steuart at Samulcottah. The regimental colour received, bore for the first time the honour "Seringapatam," which had been sanctioned in 1820.

1863 On the 17th October 1863 Orders were published that all colours issued to Native Infantry Regiments after that date should be made of silk; the dimensions to be 4 ft. flying, and 3 ft. 6 in. deep on the pike, the length of the pike, spear and ferrule included, to be 9 ft. 10 in.; the cords and tassels of the whole to be crimson and gold mixed.

The regiment received new colours, on the 12th December 1863, at Trichinopoly, which were presented by Lieutenant-General Sir J. Hope Grant, G.C.B., the Commander-in-Chief of the Madras Army, the old colours having been in use for just twenty-one years.

1867 G.O.C., No. 515, of 1867, published the following letter from the Right Honourable the Secretary of State for India:—

"INDIA OFFICE, LONDON, *8th April* 1867.

"To His Excellency the Right Honourable the Governor-General of India in Council.

"SIR,—I have the honour to inform you that the Queen has approved of a proposal to substitute a Royal Crown on the colours of Regiments of Native Infantry for the Lion and Crown—the crest of the East India Company.

"This alteration will be made as demands for new colours are received.—I have, etc.,

(Signed) "STAFFORD H. NORTHCOTE."

16

1873 The following is a copy of G.O.C.C., dated 13th December 1873 :—

"*Colours.*

"1. Under instructions from the Government, the Commander-in-Chief directs it to be notified that, as a rule, twenty years are the minimum period of duration for colours of Native Corps ; but should they at any time, before the expiration of that period, be reported by a Special Committee to have become unserviceable from extraordinary exposure, or by fair wear and tear, the special orders of Government should be obtained for their replacement.

"2. Condemned colours are to be transferred to the Ordnance Department, to be hung up in the armoury at arsenals whenever their retention is not specially allowed by His Excellency the Commander-in-Chief, to whom previous application should be made for permission to do so."

1882 In June 1882 staves and pikes for the Queen's and regimental colours were ordered to be supplied by the Clothing Department.

"The colours of Native Infantry Regiments are to be of silk ; the dimensions to be 3 ft. 9 in. flying, and 3 ft. deep on the pike, exclusive of the fringe, which is about 2 in. in depth. The length of the pike, including the royal crest, to be 8 ft. and 7½ in. ; the cords and tassels to be crimson and gold mixed.

"The royal or first colour of every regiment is to be the great union—the imperial colour of the United Kingdom of Great Britain and Ireland—in which the cross of St. George is conjoined with the crosses of St. Andrew and St. Patrick on a blue field. The first colour is to bear in the centre the imperial crown, and the number of the regiment underneath in gold Roman characters.

"The regimental or second colour is to be the colour of the facings of the regiment, except in those regiments which are faced with scarlet, white, or black ; in those regiments which are faced with scarlet or white, the second colour is to be the red cross of St. George in a white field ; in those regiments which are faced with black, the second colour is to be St. George's cross. The number of the regiment is to be embroidered in gold Roman characters in the centre.

"The regimental or second colour is also to bear the devices, distinctions, and mottoes which have been conferred by authority ; the whole to be ensigned with the imperial crown. Regiments with titles will bear such designation on a red ground round a circle within the union wreath of roses, thistles, and shamrocks."

1892 New colours were issued to the regiment on the 30th March 1892 at Bangalore. The regimental colour received was inscribed for the first time with the honours "Carnatic," "Sholinghur," "Mysore," and "Burma, 1885–87," which had been sanctioned in 1889 and 1891 in addition to the word "Seringapatam" previously sanctioned.[1] To use the language of a former Governor of Madras: "These colours, while exhibiting a proud memorial of past achievements, should never cease to wave over soldiers whose good conduct in garrison and bravery in the field should well maintain what has been so nobly won by their predecessors in arms."

[1] Application was made to retain the old colours, which was sanctioned; and they are still in the possession of the regiment.

PART VI

(1) ARMS, AMMUNITION, AND ACCOUTREMENTS.

(2) TABLE OF RESULTS OF ANNUAL COURSES.

(3) SHOOTING PRIZES.

VI

(1) ALTERATIONS IN THE ARMS AND ACCOUTREMENTS OF THE REGIMENT

1776

WHEN the 13th Carnatic Battalion came into existence, matchlocks, bows and arrows, spears, bucklers, daggers, etc., which were the arms the first sepoy levies received, were already out of date, and the regiment was armed from the first with flint-locked muskets, commonly called "Brown Bess." The weight of the musket with bayonet was 11 lb. 4 oz., the length of the barrel 3 ft. 3 in., the diameter of the bore ·753 in., and charge of powder 6 drms., with 3 flints to every 60 rounds. The priming was put into the pan from a flask containing a fine grained powder. This old musket was heavy, certainly, but when it reached its billet, the bullet from "Old Brown Bess" made a hole or smashed a bone in a manner which quickly stopped the progress of the enemy. It would appear, however, from an old inspection return of a native regiment, that the "firings" were often attended with danger, for on a certain occasion a sepoy's piece burst and wounded three men. The length of the "Brown Bess" bayonet was 17 inches. The leather accoutrements consisted of pouches, belts and frogs, and bayonet scabbards. Government authorised Commanding Officers to furnish indents, countersigned by the Commanding Officer on the spot, for whatever ammunition (powder, lead, cartridge paper, etc.) might be necessary for exercising the sepoys during the year.

1777 The European officers, the native commandant, and the company officers of the grenadier companies were, in July 1777, ordered to carry fusils in addition to swords, instead of the heavy common firelocks which they had previously carried, and all other native officers to carry spontoons in place of swords. All muskets embracing the combination of the action of the wheel-lock and the snaphaunce (as represented by the flint-lock) obtained the name of "fusils," a French adaptation of the Italian word "focile," a flint.[1] The spontoon was a weapon bearing resemblance to a halberd, and was a medium for signalling orders to the regiment. In the evolutions of the infantry, when an officer "planted" his spontoon in the ground, the men halted; when he pointed it forwards, they advanced; when backwards, they retreated. It would appear that the officers saluted with their spontoons and pulled off their hats when passing the General. "Doubtless, the salute with the sword and left hand to the shako, in the march-past in slow time of the last military generation, had emerged out of the more ancient salute with the half pike and hat."

The sergeants were armed with swords, and each carried a rattan in his hand.

1786 In 1786 the fusils and spontoons carried by the officers were superseded by swords.

1801 The following extract from G.O.C.C., dated 14th September 1801, is interesting in these days:—

"In order to obviate the inconvenience that has occurred since the introduction of the regulations for mounting guards, from the want of uniformity in the method of inspecting arms, etc., as practised by different corps, the Commander-in-Chief is pleased to establish the following mode for the inspection of arms, ammunition, and small stores; and to order that the forces serving under the Presidency of Fort St. George shall strictly adhere to it.

"The soldier standing at ease with ordered arms — attention — fix

[1] "Fusiliers" were foot-soldiers formerly armed with *fusees* with slings to sling them.

bayonets—shoulder arms—open pan—port arms—shoulder arms—shut pan—draw ramrod. In performing this motion, the soldier is to keep his proper front; the ramrod to be drawn and dropped into the barrel, and the soldier's right hand to quit the firelock.

"Upon the inspecting officer's approach, the ramrod is to be seized and sprung into the barrel, to be caught in the middle, drawn out of the barrel, and laid across the soldier's right shoulder—head of the ramrod to the front. When the officer has passed, the ramrod to be returned by the soldier, who will stand ready for the next word of command—order arms—to the right half face—stand at ease. In doing which the muzzle of the firelock is to be dropped across the body into the hollow of the left arm, the butt being kept fast. The pouch is to be opened with the right hand, and the lower part of the flap seized with the left hand. The small stores to be held in the right hand, palm to the front, and elbow raised. The ammunition and small stores having been inspected, and the officer passed, the soldier, after buttoning his pouch, is to bring his firelock back to the position of 'ordered arms,' coming to his proper front and waiting for the word of command—attention—unfix bayonets—shoulder arms."

1812

On the 2nd March 1812 it was ordered that the sergeants and havildars of native infantry should be armed in future with halberds and swords slung in frog belts. This halberd was a weapon consisting of a strong wooden shaft about six feet in length, surmounted by an instrument much resembling a bill-hook, constructed alike for cutting and thrusting, with a cross piece of steel, less sharp, for the purpose of pushing; one end of this cross piece was turned down as a hook, for use in tearing down works against which an attack was made.

At the same time, orders were issued for the sergeants and havildars of light companies to carry fusils and small fusil cartouch boxes. The men in the light companies were armed and equipped in such a manner as to allow them greater freedom of action than the rest of the regiment, and were usually employed as skirmishers to cover the advance or retreat of the line. Both the officers and havildars of the light companies carried bayonets as well as fusils, and wore cross shoulder belts. The men were armed with short hand muskets, and some of them appear to have carried hatchets suspended from the belt. The fusils carried a ball of five ounces.

In the same year company letters were introduced,

and the arms and accoutrements marked with the letter of the company and number of the man.

1816 On the 5th January 1816 it was ordered that "for the future blank cartridges would be made up in blue paper, and ball in brown paper."

1817 Regulations for sword exercise were first introduced in 1817, and on the 10th October of that year the following Order was issued:—"The Commander-in-Chief is pleased to direct that . . . one native officer and three non-commissioned officers from each battalion of native infantry within the frontiers shall be immediately detached to Poonamallee to be instructed in the infantry sword exercise lately established."

1818 The muskets were first ordered to be browned in General Orders of the 16th October 1818.

1821 In 1821 it was ordered that blacking should be used to clean the pouches and black accoutrements in place of heel ball.

1832 On the 6th December 1832 field officers were directed to wear brass scabbards instead of leather or steel scabbards.

1840 On the 20th November 1840 orders were issued for "the drums to have the crown, and number of the regiment under it, painted upon the front."

1851 On the 23rd June 1851 the regiment was armed with percussion smooth-bore muskets, in place of the flint-locks with which up to this time it had been armed (D.A.A.G.'s C Dn. letter, No. 118, dated 23rd May 1851). The new musket had a block or backsight for 150 yards. The weight of the musket and bayonet was 11 lb. 6 oz., the length of the barrel and size of the bore was the same as the flint-lock, but the charge of the powder was reduced to 4½ drms. The bore of the British musket, being larger than that of other countries, was considered an advantage, because their balls could be fired out of our barrels, whilst our balls could not be fired out of theirs. It was also thought that the greater weight of the British ball produced an increased range and momentum; this was, however, counteracted by the excess of windage.

By G.O.G., No. 42, of 28th February 1851, waist-belts were introduced in lieu of the shoulder belts from which the bayonets were previously suspended, but the regiment did not receive these until November 1856. At the same time packs were introduced instead of knapsacks, to the great delight of the sepoy.

1855 The allotment of shotted cartridges for practice to Native Infantry Regiments was increased from 36 to 60 shotted cartridges, with percussion caps, according to the regulated pattern, by an Order issued on the 23rd February 1855.

1856 New pouches (cartridge) were issued to the regiment in 1856, and new pouches (cap) in June 1857. 1857

On the 5th September 1857 orders were issued for swords and sling belts to be served out to bandsmen, in lieu of muskets, belts, and pouches with which they were previously armed. They were, however, to be instructed in musketry as other soldiers.

1859 New waist-plates were received on the 3rd June 1859, and were issued to the men.

1862 New pouches were again served out to the regiment on the 17th September 1862.

On the 10th January 1862, 50 rounds of ball ammunition and 100 rounds of blank ammunition per man, with $\frac{1}{10}$th percussion caps in excess, was sanctioned annually for Native Infantry Regiments.

1864 On the 18th August 1864, Government sanctioned the issue of field bugles "of an improved construction" for Native Infantry Regiments.

1865 On the 5th July 1865, Azemar's silent drums were authorised for issue to Native Infantry Regiments, at the rate of one per two companies.

1866 New muskets of the Enfield pattern were received on the 27th November 1866 in Cannanore, and replaced the percussion smooth-bore muskets. The weight of the rifle musket with bayonet was 9 lb. 3 oz., length of barrel 3 ft. 3 in., diameter of bore .577 in., and charge of powder 2½ drms. The number of grooves was three, having one turn in 78 inches. The bayonet was secured with a locking ring. The barrel was secured to the stock by three steel bands fastened by screws.

1867 On the 13th June 1867 it was ordered that the larger pouch should hang four fingers below the elbow, the small pouch to be worn in front of the right side.

On the 1st July 1867 steel scabbards were ordered to be worn by all European officers of Indian Regiments under the rank of major, instead of the leather sword scabbards.

1868 On the 7th May 1868, 50 rounds of ball ammunition, 100 rounds of blank ammunition, with $\frac{1}{16}$th spare percussion caps in excess of the number of rounds allowed, were fixed for each man's annual practice.

On the 9th December 1868 buff ball bags, similar to those then supplied to European troops, were substituted for the twenty-round black leather pouches then in use.

1870 On the 29th September 1870 the regiment was rearmed with new Enfield rifles at Cannanore, previous to embarking for foreign service in Hong-Kong.

The composition for greasing the cartridges of the Enfield rifles consisted of eight parts of gingelly oil and seven parts of beeswax.

As these cartridges were usually kept in parcels of ten each, and as, after firing a few rounds, the remainder, if left loose in the pouch, were very liable to injury by being shaken, it was ordered that they should always, if possible, be tied together, and that this should be done "before moving even the shortest distance from the field or practice ground."

Early in 1872 a bass drum was allowed to each Native Infantry Regiment. 1872

On the 14th November 1872 50-round pouches were authorised, in lieu of the 40-round pouches, as the native army had been armed with Enfield rifles.

On the 3rd December 1872 regiments armed with the Enfield rifles were ordered to be supplied with "wrenches, nipple with cramp," at the rate of one for each havildar and naick; and "wrenches, nipple without cramp," at the rate of one for each sepoy.

One arm chest per company was also sanctioned.

On the 7th January 1874 runners and loops were sanctioned for the rifle slings. 1874

On the 18th April 1874 trigger testers were sanctioned.

A new pattern oil-bottle, with the screw inside the cap, designed to prevent leakage, was ordered to be taken into use on the 11th October 1874.

In order to prevent the loss of ammunition from the expense pouches and ball bags of the men when skirmishing in the field, the addition of serrated flaps to all ball bags and expense pouches was directed on the 18th February 1876. Regiments of native infantry in possession of 10-round ball bags were ordered to alter their pouches. 1876

On the 10th August 1876 10 rounds extra ball ammunition were sanctioned for "field-firing."

On the 8th February 1877 the issue gratis of 6 rounds of ball ammunition per man was sanctioned for "match shooting" in Native Infantry Regiments armed with muzzle-loading rifles, and 10 rounds per man in regiments armed with Sniders. 1877

On the 12th May 1877 new Snider rifles were issued to the regiment at Madras. A short rifle with sword-bayonet was served out to each havildar.

The old pattern bayonets (length 17½ in.), which had

been in use with the muzzle-loading rifles, were retained by the men for the Sniders.

1878 By S.G.O., No. 25, of 1878, the band instruments for Native Infantry Regiments were ordered to be obtained through the Commandant, Kneller Hall School of Music, on the same terms as supplied to British corps.

By S.G.O., No. 31, of 1878, a scale of materials was authorised to be supplied biennially for browning arms of native corps, in which were included three pints of Rangoon oil per 100 stand of arms.

On the 25th June 1878 the following instructions for packing ammunition in the pouches were published :—

"*Pouches of rank and file.*—In cantonments, 20 rounds to be placed across the small compartment on the left, alternately base and point. On the line of march, 30 rounds to be placed alternately, base and point, across the large compartment, reversing the tin, from which the cap magazine should be removed, and 10 rounds in the small bag.

"*Havildars' Pouches.*—The entire partition to be removed so as to admit of 20 rounds being packed therein. Should there be any difficulty in buttoning the pouch when packed with 50 rounds, the strap is to be lowered to the extent required, or the hole therein enlarged."

1879 By G.O.C.C., No. 44, of 1879, the substitution of brown for buff leather accoutrements was sanctioned for the Madras Native Army, the change to take place as the buff leather became worn out.

It was not, however, until the 22nd October 1882, when the regiment was in Jubbulpur, that the brown leather accoutrements (valise equipment) were issued to the men, in place of the buff leather accoutrements then in use. The British and native officers also took brown leather belts and sword knots into use, as it had been ordered on the 10th February 1880, that brown leather belts and sword knots should be worn by both British and native officers when the men were served out with brown leather accoutrements. The native officers received them free of charge. Shortly afterwards the British officers were directed to provide themselves with sabretaches of brown leather.

The alteration of ammunition bags in possession of Native Infantry Regiments, by the insertion of a larger flap, was sanctioned on the 31st July 1881. 1881

Revolvers were first sanctioned as part of all officers' equipment by I.A.C., Clause 142, of July 1881, and the native officers received them the same year at Jubbulpur. These were the Enfield pattern. One revolver with leather case and pouch was served out to each native officer, and an annual allowance of 24 rounds of revolver ammunition was sanctioned.

The abolition in the Indian Army of the infantry officer's sabretache was ordered on the 21st June 1883. 1883

On the 14th July 1883 an annual allowance of 60 rounds of ball ammunition for each officer, 130 rounds of ball ammunition (with 10 rounds extra if battalion field-firing was carried out), and 100 rounds of blank ammunition per trained soldier, was sanctioned, as well as 93 rounds of ball ammunition and 40 rounds of blank ammunition per recruit.

On the 30th September 1890 the issue of revolvers (and 24 rounds of revolver ammunition per annum) was sanctioned to the hospital assistants. 1890

Clause 172 of I.A.C. of November 1888 having sanctioned signalling equipment for one Native Infantry Regiment at Bangalore, and the 13th Madras Infantry having been detailed on mobilisation for the 1st Army Corps in 1890 after arrival at that station, a complete set was then issued to the regiment for the first time. This equipment was not to be taken away by the regiment moving in relief, but in case of mobilisation was to form part of its field service stores. It consisted of 16 flags of various colours and sizes, 2 heliographs (5 in.), 2 telescopes, 2 hand-signalling lamps, 2 Begbie's B.B. pattern lamps, 2 signalling cypher wheels, 2 dummy keys, etc. (as subsequently authorised by Clause 96 of I.A.C., 1893).

Clause 173 of I.A.C. of November 1888 sanctioned the following scale of light entrenching tools for issue to corps at certain mobilisation stations, for use on field service. They were ordered to be preserved for employment on service only, and were received by the regiment in 1890 from the 25th Madras Infantry :—

Axes, pick, helved, 2¾ lb.		90
Bars, crow,	5 ft. 6 in.	2
	4 ft. 6 in.	2
Hammers, sledge, 10 lb.		2
Hooks, bill, handled		40
Shovels, helved, cast-steel		90

These tools were ordered to be carried on five pairs of kajawahs (iron, mule), which were to be used in peace time for carrying the ordinary picks and shovels allowed for trench drill.

1891 The free issue of swords to native officers was discontinued in January 1891, after which date they were issued by the Arsenal on payment, and became their own property.

In May 1891 the regiment received the "Mackenzie equipment" at Bangalore in lieu of the valise equipment. The Mackenzie pattern belt consists of a broad leather waist-belt in three pieces, with plain brass waist-plate in front, and buckles on the right and left side.

Attached to it is the usual frog attachment for carrying the sword-bayonet, and three ammunition pouches for 40 rounds each, two of which are carried in front by means of loops through which the waist-belt passes. Braces attached behind to D's in the waist-belt, crossed behind like ordinary braces, are fixed in front by buckles to D's on the pouch, and take the weight of the ammunition off the man's stomach. The third pouch is carried on the waist-belt behind. The greatcoat is carried on the back by two straps which pass through loops on the braces.

In June 1891 the corps was selected for re-armament with Martini-Henry rifles, Mark IV., and sword-bayonets; the old Sniders and bayonets being returned to the Madras Arsenal.

The rifles and sword-bayonets issued to the men were the same as those issued to the havildars. The usual proportion of long and short butt rifles were served out. The majority of the sword-bayonets were stamped with the name of "Wilkinson." The frogs for the triangular bayonets were altered to suit them to the new sword-bayonet. Two pouches (40 rounds) were served out to each man.

1894

In 1894 the drummers' swords were returned to the Arsenal, and replaced by sword-bayonets and frogs as worn by the rank and file.

Ten thousand rounds of smokeless powder having been issued to the regiment for experimental purposes, several interesting trials took place, a full report of which was submitted to the A.A.G. for Musketry.

In November 1894, Webley's Mark III (Service Model) revolvers were received and issued to native officers and hospital assistants, in supercession of the Enfields, which until then had been in use.

In November 1894 new pouches (20 rounds) at the rate of two per man were received, one of the 40-rounds pouches per man being returned to the Madras Arsenal. Thus each man has now one 40-rounds pouch and two 20-rounds pouches.

1896

By Army Orders of the 1st January 1896, officers were directed to provide themselves with a new pattern sword, or have those in their possession rehilted. The new sword had a steel half-basket hilt, pierced with scroll design and royal cypher (V.R.) and crown chases. At the same time all mounted officers were ordered to wear steel scabbards and steel spurs of swan-neck pattern, except in the evening and at levees, when brass spurs were to be worn. Thus the brass spur of swan-neck pattern, and the brass scabbard, till then worn by field and staff officers, were abolished.

Permission having been granted on the 7th March 1892 for the wearing of "Sam Browne" belts with slings (as laid down for dismounted officers) by native officers, they were taken into wear by all the native officers of the regiment on the 1st April 1896.

1897 By G.O.C., No. 809, dated the 25th November 1896, the Commander-in-Chief was pleased to decide that the royal and imperial cypher (V.R.I.) below the crown should be retained as the device on the new steel sword-hilt by officers of the Indian Army.

In 1897 revolver ammunition was sanctioned annually to each British and native officer, in lieu of the 60 rounds of rifle ammunition.

(2) TABLE OF RESULTS OF ANNUAL COURSES.

The following abstracts of Ball Practice of the regiment for the years noted are interesting :—

Year.	96 paces.	100 paces.	120 paces.	142 paces.	150 paces.	180 paces.	240 paces.
1842		1 in 3[1]	1 in 2½		1 in 4½	...	...
1844	1 in 2¾	...	1 in 4	1 in 5½	...	...	...
1845	1 in 3½	...	1 in 4¼	1 in 5¼	...	...	...
1846	1 in 3	...	1 in 4	1 in 5	...	1 in 8¼	1 in 15¼
1847	1 in 2¾	...	1 in 3¼	1 in 5	...	1 in 8½	...
1848	1 in 2½	...	1 in 3	1 in 4½	...	...	...
1849	1 in 3¼	...	1 in 3½	1 in 4½	...	1 in 7	...

[1] That is, one hit in 3 shots (average).

The following is an interesting table as comparing the Average Ranges of the Musket, Enfield, and Martini-Henry rifle :—

	Regiment fired annual courses.		
	Musket (1776 to 1872).	*Enfield* (1872 to 1891).	*Martini-Henry* (1891 to date).
Accurate fire	100 yards	600 yards	1200 yards
Effective against detached parties	150 ,,	800 ,,	1500 ,,
Effective against troops in column	200 ,,	1000 ,,	1800 ,,

Table of Results of Annual Courses since 1869–70.

Year.	Weapon.	No. on List of Madras Infantry Regiments.	No. on List of N.I. Regiments of India.	Figure of Merit.	Standard of Merit.	Percentage 1st Per. Indiv.	Percentage 2nd Per. Collec.	Percentage Field Practices.	Best Shooting Company in Regiment.	Best Shot in Regiment.
1869–70	Percussion smooth-bore muskets	5	...	47.42	...	...	...	...	No. 7 / 50.90	
1871–72		...	...	54.19	...	29.79	...	...	...	
1872–73	Enfield rifles	8	...	...	...	...	...	...	...	
1873–74	,,	N.E.	...	...	...	...	...	...	...	
1874–75	,,	27	...	52.87	...	...	...	...	A / 64.58	Naick Vurdiah
1875–76	,,	28	...	54.97	...	...	...	...	H / 60.29	Private Ballagooro
1876–77	,,	29	...	...	...	...	...	...	...	
1877–78	Snider rifles	20	...	...	...	...	...	...	...	
1878–79	,,	...	43 Bengal	64.29	...	...	...	...	H / 74.35	No. 291, Private Sunthiah
1879–80	,,	...	...	56.50	...	...	...	...	D / 65.51	No. 539, Havildar Sashachellum
1880–81	,,	...	...	62.56	...	...	...	...	B / 70.83	Naick Veeragooloo
1881–82	,,	...	...	76.17	...	...	...	...	D / 88.54	Naick Chendanum
1882–83	,,	N.E.	...	...	...	...	...	...	...	
1883–84	,,	N.E.	...	...	...	...	...	...	...	
1884–85	,,	N.E.	...	...	...	...	...	...	...	
1885–86	,,	8	49	111.17	Moderate	...	...	...	...	
1886–87	,,	N.E.	On Service.							
1887–88	,,	N.E.								
1888–89	,,	N.E.								
1889–90	,,	20	70	68.66	Moderate	40.40	28.26	...	G	
1890–91	,,	21	74	69.26	Bad	...	...	35.07	D	No. 1462, Naick Muhammad Ameer
1891–92	Martini-Henry rifle, Mk. IV.	7	65	38.46	Moderate	46.16	38.46	45.18	D / 51.18	No. 1773, Pte. Muhammad Mustafa, 118 points
1892–93	,,	4	51	47.23	Good	51.47	47.23	38.90	E / 51.31	No. 1462, Naick Muhammad Ameer, 122 points
1893–94	,,	8	74	47.87	,,	51.91	47.87	43.90	F / 55.74	No. 1079, Havildar Razack Ali Beg, 117 points
1894–95	,,	3	29	55.82	Very good	52.94	55.82	53.69	D / 65.92	No. 1462, Naick Muhammad Ameer, 119 points
1895–96	,,	12	...	53	Good	50	52	53	D / 59	No. 1636, Lce-Naick Surinarayudu, 109 points
1896–97	,,	13	...	54	,,	53 Mod.	54	56	D / 59	No. 1844, Lce-Naick Ramanna, 106 pts.

N.E. = Not exercised.

(3) PRIZES FOR SHOOTING.

Extract from G.O.C.C., No. 194, dated 17th April 1884:— 1884

"ADJUTANT-GENERAL'S OFFICE,
Simla, 12*th March* 1884.

"It is notified for information that the best shot in the Native Armies of the three Presidencies, and the winner of the first, or army prize, of Rs. 100, with a silver medal, is No. 578, Havildar Cunniah, 13th Madras Native Infantry."

No. 22, Drummer Davies, was awarded a silver medal and Rs.25 (third prize) in the Native Championship of the S.I.R.A. Meeting, held at Bangalore in September 1891. 1891

In September 1891, No. 1971, Havildar Abdur Rahim obtained the second prize in the Bengal Presidency Rifle Association competition at the S.I.R.A. Meeting, and was presented with a silver medal and Rs.35.

The regiment was very successful at the Southern India Rifle Association Meeting held at Bangalore in September 1892, winning no less than Rs.1030, 12a. in cash, and six first prizes. 1892

The following is a list of some of the principal prizes won at that meeting:—

The *Cubbon Cup*, value Rs.600, in the competition for which twelve Madras Regiments competed. The regimental team was composed of the following officers and non-commissioned officers:—Lieutenant H. R. Baker, Subadar Kadir Moideen, Havildar Razack Ali Beg, Havildar Sheikh Emam, Havildar Abdur Rahim, Naick Muhammad Ameer, Lance-Naick Muhammad Mustafa, and Lance-Naick Abdul Curreem. The best shot in the winning team was Havildar Razack Ali Beg.

The *S.I.R.A. Silver Medal* and Rs.50, being the Native Champion Plate, was won by Naick Poonoosawmy of the regiment, who was also successful in carrying off several first and other prizes during the meeting.

The *Commander-in-Chief's Prize.*—The first and second prizes were divided equally between the 13th and 17th Madras Infantry.

Teams from the regiment obtained the second prize in the *Dewan's prize* and also in the *Assaye prize.*

1894 The native champion at the S.I.R.A. Meeting, held at Bangalore in September 1894, was Havildar, No. 1971, Abdur Rahim, who was presented with a silver medal and Rs.50.

Private, No. 1204, Rungasawmi was awarded the 3rd prize of the Native Championship at the above meeting, which consisted of a silver medal and Rs.25.

1896 The regiment was very successful at the fifth annual meeting of the Burma Rifle Meeting, held at Mandalay in December 1896, winning about Rs.600 in cash, a match-shooting revolver, a hunting saddle, and six first prizes.

The *Inter-regimental Revolver Cup* was won by the 13th Madras Infantry. The team, composed of Lieutenants R. P. Jackson, W. C. S. Prince, H. R. Baker, and W. C. Anderson, made 426 points, of which Lieutenant W. C. S. Prince contributed 115—the highest individual score in the competition—thereby winning the match revolver presented by Messrs. J. Rigby & Co., London.

Havildar Venketsami won the medal and Rs.20 awarded by the S.I.R.A., being the second prize for the championship.

PART VII

(1) LIST OF INSPECTING OFFICERS.

(2) SUCCESSION OF BRITISH REGIMENTAL STAFF OFFICERS.

(3) LIST OF BRITISH REGIMENTAL OFFICERS.

(4) LIST OF OFFICERS KILLED AND WOUNDED.

(5) LIST OF OFFICERS AND OTHERS WHO HAVE DISTINGUISHED THEMSELVES.

(6) LIST OF NATIVE OFFICERS.

(7) LIST OF RECIPIENTS OF THE ORDER OF BRITISH INDIA.

(8) LIST OF RECIPIENTS OF GOOD CONDUCT MEDALS.

(9) RETURN OF EFFICIENCY OF SIGNALLERS.

(10) RETURN OF CERTIFICATES OF EDUCATION.

(11) MEDICAL STATISTICS.

VII

(1) LIST OF OFFICERS WHO HAVE INSPECTED THE REGIMENT.[1]

Major-General J. Floyd	Jan. 29, 1799	Dindigul
Major-General M'Leod commanding Ceded Districts	Dec. 19, 1865	Cannanore
Brigadier-General De Sausmarez . .	Jan. 15, 1866	"
" "	Jan. 23, 1867	"
Major-General W. K. Babington . .	Jan. 8, 1868	"
Major-General C. B. Hodson . .	Dec. 15, 1868	"
Brigadier-General G. Selby . . .	Jan. 24, 1870	"
Major-General Whitfield . . .	Mar. 20, 1871	Hong-Kong
" "	Feb. 2, 1872	"
Brigadier-General G. B. Shakespear .	Mar. 21, 1873	Madras
" "	Dec. 12, 1873	"
Brigadier-General T. Raikes, C.B. .	Jan. 19, 1876	"
" "	Jan. 29, 1877	"
" "	Dec. 3, 1877	"
Major-General T. Raikes, C.B. . .	Nov. 11, 1878	"
Brigadier-General I. Murray, C.B. .	Mar. 4, 1880	Meean-Meer
Brigadier-General A. H. Cobbe, C.B. .	Mar. 22, 1881	Jubbulpur
H.E. Lieut.-General Sir Fred. S. Roberts, G.C.B., C.I.E., V.C.	Feb. 25, 1882	"
Brigadier-General A. H. Cobbe, C.B. .	Feb. 27, 1882	"
Brig.-General H. C. Wilkinson, C.B. .	Mar. 16, 1883	"
H.E. Lieut.-General Sir F. S. Roberts, G.C.B., C.I.E., V.C.	Feb. 15, 1884	"
Brigadier-General A. H. Murray, C.B. .	Feb. 22, 1884	"
Brigadier-General W. A. Gib, C.B. .	Mar. 16, 1885	Bellary
Brigadier-General H. H. Foord . .	Feb. 16, 1886	"
Brigadier-General H. Collett, C.B. .	Oct. 17, 1888	Pyawbwé
Brigadier-General Gordon . . .	1889	Meiktila

[1] The system of periodical inspection and review of regiments dates from 5th Apri 1797, before which time systematical inspections were not insisted upon.

Brigadier-General G. J. Smart . .	Mar. 5. 1890	Pallaveram
Lieut.-General Sir Charles Arbuthnot, K.C.B.	Oct. 10, 1890	Bangalore
Lieutenant-General Sir James Dormer .	Mar. 31, 1891	,,
Brig.-General H. M. Bengough, C.B. .	Feb. 20, 1891	,,
Brigadier-General E. Faunce, C.B. .	Mar. 7, 1892	,,
H.E. the Commander-in-Chief, Madras (Lieut.-General Sir James Dormer)	Sept. 28, 1892	,,
H.E. the Viceroy (Lord Lansdowne) . .	Nov. 23, 1892	,,
Brigadier-General G. Rowlandson .	Jan. 20, 1893	Cannanore
Major-General C. J. East, C.B. (Provisional Commander-in-Chief) . .	Sept. 20, 1893	,,
Brigadier-General G. Rowlandson .	Jan. 17, 1894	,,
H.E. Lieutenant-General C. Mansfield Clarke, C.B.	Sept. 25, 1894	,,
Brigadier-General J. W. Prendergast .	Oct. 22, 1894	,,
,, ,,	Jan. 23, 1895	,,
H.E. Lieutenant-General C. Mansfield Clarke, C.B.	Sept. 10, 1895	,,
Major-General E. Stedman, C.B. . .	Dec. 19, 1896	Thayetmyo
H.E. Lieutenant-General C. Mansfield Clarke, C.B.	Feb. 5, 1896	,,
Brigadier-General J. T. Cummins, D.S.O.	Feb. 27, 1896	,,
,, ,,	Dec. 4, 1896	,,
H.E. Lieutenant-General Sir C. Mansfield Clarke, K.C.B.	Jan. 8, 1897	,,
Brigadier-General Black commanding . Nagpore District	Feb. 7, 1898	Raipur

(2) SUCCESSION OF BRITISH REGIMENTAL STAFF OFFICERS.

Colonels.		Commanding Officers.		Adjutants.		Quartermasters (Interpreters and Paymasters).		Surgeons.		Assistant Surgeons.	
		Capt. Hugh Robert Alcock	12/1776	Lieut. Richard Anderson	12/1776	*From Dec. 1776 to 25 June 1800 the duties of Quartermaster were performed by the Adjutant, and, after the latter date, by a sergeant*		*From Dec. 1776 to Dec. 1785—No person in medical charge. From Dec. 1785 to July 1796, native doctors, who were paid partly by Government and partly by the native ranks, were attached to the regiment.*			
		Capt. George Clarke	6 4/1783	Lieut. George Wahab	1778						
		Capt. James Oliver	2/11/1783								
A Colonel first appointed	12/7/1796	Major Colin Campbell (offg.)	1799	Lieut. Peter H. Keay	23/1/1787						
				Lieut. Charles Rand	10/1798			*A surgeon for both battalions first appointed*	12/7/1796	Inverarity . .	12/7/1796
Edward Collins	12/7/1796	Lieut.-Col. John Chalmers	31/7/1799	Lieut. Thomas Baynes	1798	*A subaltern first appointed Quartermaster and Interpreter*	7/1817				
Urban Vigors .	1/1/1800	Lieut.-Col. William Kenny	1/1/1800	Lieut. Charles Lucas	1799			James Barter .	17/10/1800		
				Lieut. George Sinclair	1/1/1800						
		Major John Kennet	13/1/1801								
		Lieut.-Col. Edward Gibbings	14/1/1801	Lieut. Arthur Molesworth	1800						
		Lieut.-Col. James Innes	5/3/1801	Lieut. Francis Evans	27/8/1800			William Colhown	14/4/1801		
William Kinsey	1/1/1803	Capt. Charles Lucas (offg.)	1803	Lieut. John William White	2/6/1801					B. Humpage .	15/9/1802
										A. Inverarity .	1802
		Lieut.-Col. Haliburton	1804	Lieut. Thomas T. Stevenson	14/10/1804					John Harlem .	24/1/1803
		Lieut.-Col. John Cuppage	30/5/1804								
		Lieut-Col. R. Warne	31/5/1804								
		Lieut.-Col. J. Long	1805	Lieut. Henry John Wilkinson	5/7/1805						
		Lieut.-Col. Lindsay	12/12/1805								
		Lieut.-Col. Alexander Baillie	13/12/1805								
		Lieutenant-Colonel M'Kerras	1806								
		Lieut.-Col. Poole H. Vesey	21/6/1807							Thomas Sutton .	3/2/1808
		Lieut.-Col. Corner	12/10/1808	Lieut. George Hunter	12/1/1808					W. S. Anderson .	3/10/1808

SUCCESSION OF BRITISH REGIMENTAL STAFF OFFICERS—*coutinued*

Colonels.		Commanding Officers.		Adjutants.		Quartermasters (Interpreters and Paymasters).		Surgeons.		Assistant Surgeons.	
		Lieut.-Col. George Martin	2/1/1809	Lieut. Cadenskie (offg.)	9/8/1809						
		Lieut.-Col. Robert Barclay	15/10/1809								
		Lieut.-Col. William Orrok	9/2/1810	Lieut. Robert Inverarity	19/2/1813			John Grant . .	1810	A. Campbell . Adam Stevenson .	16/11/1812 3/9/1816
		Lieut.-Col. Henry Webber	1811								
		Lieut.-Col. M. Wilkes	1813								
		Lieut.-Col. Sam. Wm. Ogg	4/8/1816								
		Lieut.-Col. Henry Bowen	18/8/1818								
		Lieut.-Col. Thomas A. Fraser	28/1/1819								
		Lieut.-Col. Charles Deacon	25/1/1821								
		Lieut.-Col. J. Mackenzie, C.B.	6/2/1821	Lieut. James Maxtone	31/5/1819	Lieut. R. Inverarity	31/5/1819	Kenneth Macaulay	26/10/1815 15/12/1834	Thomas Edwards	25/9/1820
		Lieut.-Col. Henry H. Pepper	22/10/1822			Lieut. Thomas Dallas	9/4/1822	John Cormick .	3/9/1816		
		Lieut.-Col. C. T. G. Bishop	1823								
William M'Leod	1825	Lieut.-Col. H. F. O. Smith, C.B.	1823								
		Lieut.-Col. A. Limond	21/3/1824								
		Lieut.-Col. John Carfrae	1/5/1824								
		Lieut.-Col. George Hunter	1/6/1824	Lieut. George Dods	4/6/1824	Lieut. James Briggs	4/6/1824	Thomas Edwards	4/6/1824		

		Lieut.-Col. Fraser P. Stewart	3/6/1824	Lieut. T. G. E. G. Kenny	27/5/1823	Lieut. Charles Fladgate	1/7/1825	G. A. Herklots, M.D.	12/4/1826		
		Lieut-Col. Frederick Bowes	23/1/1826								
		Lieut.-Col. Cadell	1827								
		Lieut.-Col. B. B. Parlby, C.B.	23/1/1828								
		Lieut.-Col. Thomas King	31/12/1828			Lieut. H. C. Beevor	3/5/1829			John O'Neill .	17/10/1829
		Lieut.-Col. W. Milne	29/7/1831								
		Lieut.-Col. Josiah Stewart, C.B.	4/2/1832								
		Lieut.-Col. John Wilson	25/5/1833			Lieut. E. Slack .	22/3/1833	James Dalmahoy.	5/9/1835	H. Goodhall .	15/12/1834
		Lieut.-Col. Edward Cadogan	30/5/1833					Samuel Higginson	20/6/1836		
John Briggs .	16/11/1838	Lieut.-Col. S. W. Steel, C.B.	27/1/1840	Ensign A. Robinson	24/12/1839	Lieut. W. F. Goodwyn	14/1/1840	I. T. Maule. .	19/1/1839	A. C. B. Neill, M.D.	19/3/1842
William C. Fraser	1843	Lieut.-Col. George Dods	23/12/1842			Lieut. E. F. Burton	3/6/1845	W. Burrell . .	9/1/1845	J. Dorward . .	28/1/1843
										J. Pringle, M.D. .	8/11/1844
		Lieut.-Col. J. H. Winbolt	15/3/1845	Lieut. E. F. Burton	7/7/1846			G. A. C. Bright .	25/9/1845	A. Macintosh, M.D.	16/9/1845
										D. T. Morton .	2/3/1847
		Lieut.-Col. R. Thorpe	9/3/1848	Lieut. F. A. Brooking	24/11/1848	Lieut. C. W. Taylor	6/7/1847			A. H. Ashley .	27/11/1847
J. H. Winbolt .	1848	Lieut.-Col. George Hutton	4/10/1848							J. Cadenhead .	17/5/1848
L. W. Watson.	1849							W. Gilchrist .	1849	S. T. Lyell . .	19/2/1849
		Lieut.-Col. J. D. Stokes	3/3/1850					S. T. Lyell .	5/10/1849		
		Lieut.-Col. C. A. Browne	1851	Lieut. C. W. Taylor	21/2/1851	Lieut. J. H. Warden	21/2/1851			H. R. Oswald .	12/5/1851
										J. Brett, M.D. .	28/2/1852
		Lieut.-Col. G. C. Hughes	2/11/1854	Lieut. W. M. Williams	21/11/1854					W. Williamson .	3/3/1854
										J. M'Donald .	7/11/1855
		Lieut.-Col. T. A. Duke	1857								
John Briggs .	16/3/1858	Lieut.-Col. Daniel Hall Stevenson	12/2/1857			Lieut. B. F. Heysham	28/4/1857				

SUCCESSION OF BRITISH REGIMENTAL STAFF OFFICERS—*continued.*

Colonels.		Commanding Officers.		Adjutants.		Quartermasters (Interpreters and Paymasters).		Surgeons.		Assistant Surgeons.	
		Lieut.-Col. J. W. Goldsworthy	22/10/1859								
		Lieut-Col. M'Leod	1859								
		Lieut.-Col. J. W. G. Kenny	29/9/1860							H. E. Busteed .	16/7/1860
		Lieut.-Col. J. Hill	21/1/1861			Capt. H. A. Hare	26/11/1861				
		Lieut.-Col. T. D. T. Dyer	8/1/1862								
		Lieut.-Col. A. H. Hervey	8/1/1862	Lieut. R. A. Chadwick	15/3/1863	Capt. W. Hands	28/12/1863	L. W. Stewart .	15/10/1863	J. A. Cox, M.D. .	15/3/1863
		Lieut.-Col. J. Kitson	1864	Lieut. E. J. Watson	28/12/1863					T. G. Howell .	1863
										— Beaman . .	1863
		Col. Augustus Beecher Marsack (offg.)	1865	Lieut. W. N. Wroughton	26/10/1864	Lieut. C. B. Smith	11/7/1864			L. W. Stewart .	27/8/1863
Appointment abolished	1/11/1865			Lieut. C. M. Moberly	1/11/1865	Capt. G. E. H. Beauchamp	22/11/1865				
		Major C. Pulley .	25/10/1865								
		Col. Edmund Francis Burton	1/11/1865								
		Lieut.-Col. W. T. Money	15/3/1866	Lieut. E. L. Armstrong	1866					*Appointment abolished*	1/11/1865
		Lieut.-Col. C. P. Y. Triscott	1/4/1867	Lieut. E. Moore .	17/6/1867	Lieut. E. L. Armstrong	17/6/1867	H. J. Beach .	19/1/1867		
						Lieut. E. P. James	22/3/1869				
		Lieut.-Col. C. W. Taylor	11/1867			Capt. H. Cunningham (offg.)	30/7 1869				
		Col. A. K. C. Kennedy (offg.)	13/9/1870 9/10/1871					R. V. Power, M.D.	22/9/1870		
		Col. Thomas Greenaway	27/7/1874								

		Col. William Alexander Riach	2/10/1875								
		Col. G. T. Hilliard (offg.)	10/1/1875 8/1/1876					J. G. Collis . .	3/5/1876		
		Col. Francis Dawson (offg.)	11/9/1876 10/11/1876					Surg.-Maj. A. H. Beaman	11/11/1876		
		Col. Augustus Beecher Marsack	29/6/1877					Surg.-Maj. J. S. Riding	5/1/1878		
		Col. G. T. Hilliard	12/9/1878					Surg. A. H. Leapingwell	12/8/1878		
		Col. Thomas Spence Hawks	27/6/1877 20/3/1880					Surg. A. H. Williams	11/10/1879		
		Col. Edmund Frederic Waterman	26/4/1879								
		Col. R. S. Couchman	17/6/1879								
		Col. W. E. White	17/12/1879								
		Lieut.-Col. William George M. Strickland	29/6/1877 20/3/1880								
		Col. William Anderson	19/9/1886	Lieut. G. A. Welman	10/8/1882	Lieut. J. E. Preston	7/6/1882	Surg.-Maj. W. J. Busteed	17/1/1882		
						Lieut. J. H. Smith	1/5/1885				
		Lieut.-Col. William Lancaster Ranking	16/4/1888	Lieut. E. W. Carrick	17/1/1890	Lieut. J. P. James	25/1/1887	Surg. A. G. E. Newland	8/11/1882		
				Lieut. R. P. Jackson	10/8/1891	Lieut. R. P. Jackson	10/8/1890	Surg. N. Chatterjie	23/11/1882		
		Lieut.-Col. Charles Henry Shepperd	8/5/1893			Lieut. W. C. S. Prince	10/8/1891				
		Lieut.-Col. George Carew Fenwick	12/6/1893								
		Captain Herbert Lawson	26/5/1894								

(3) LIST OF BRITISH OFFICERS WHO HAVE SERVED IN THE REGIMENT

(In Order of Joining)

Names.	Appointments with Dates.		Remarks with Dates.	
Hugh Robert Alcock	Captain . .	4/5/1768	Commandant .	12/1776
			Transferred . .	6/4/1783
	Lieutenant-Colonel	29/5/1783	Resigned . .	5/1784
Richard Anderson .	Lieutenant . .	...	Adjutant . .	12/1776
George Wahab .	Ensign . . .	9/11/1770	Adjutant . .	1778
	Lieutenant . .	28/7/1775	Transferred to 14th Carnatic Bn.	
Caleb Pearson . .	Ensign . . .	10/12/1770		
	Lieutenant . .	26/8/1776		
John Spencer . .	Ensign . . .	18/8/1773		
George Aubery .	Ensign . . .	9/9/1773		
Thomas Sampson .	Ensign . . .	5/6/1775		
Andrew Boswell .	Ensign . . .	20/9/1775		
Thomas Davies .	Ensign . . .	21/8/1776	Died . . .	...
Thomas Phillips .	Ensign . . .	10/1/1778		
William Collins .	Ensign . . .	11/5/1778		
Harcourt Powell .	Ensign . . .	4/8/1778		
William Wemyss .	Ensign . . .	11/8/1778		
Samuel Godfrey .	Ensign . . .	25/8/1778		
William Fenwick .	Ensign . . .	27/8/1778		
George Clarke . .	Captain . .	29/12/1777	Commandant .	6/4/1783
			Transferred . .	2/11/1783
James Oliver . .	Ensign . . .	22/11/1770	Commandant .	2/11/1783
	Lieutenant . .	5/2/1776	Dismissed . .	21/6/1807
	Captain . .	2/11/1783		
	Major . . .	1/3/1794		
	Lieutenant-Colonel	1/6/1796		
Peter H. Keay . .	Lieutenant . .	23/1/1787	Adjutant . .	23/1/1787
Robert Gahagan .	Ensign . . .	14/8/1788	Died . . .	1792
	Lieutenant . .	2/3/1783		
—— Lind . .	Lieutenant . .	1782		
—— Haywood .	Ensign . . .	1782		
S. Cuppage . .	Captain . .	1796		
E. M. Gepp . .	Captain . .	1796		
—— Cunningham .	Lieutenant . .	1796		
W. Burke .	Lieutenant . .	1796		
Charles Rand . .	Lieutenant . .	1796	1st Battalion	
	Captain . .	1/1/1800	Adjt. and Q.M., 2nd Bn.	10/1798
H. Smith . .	Lieutenant . .	1796		
H. M'Intosh . .	Lieutenant . .	1796		
P. Baynes . .	Lieutenant . .	1796		
H. F. Smith . .	Lieutenant . .	1796		
S. Fish . . .	Lieutenant . .	1796		
F. Ahmutty . .	Lieutenant . .	1796		
—— Inverarity .	Assistant-Surgeon .	1796	Transferred . .	5/9/1802
Charles Aldridge .	Lieutenant . .	11/12/1793	1st Battalion	
	Captain-Lieutenant	17/6/1800		
	Captain . .	14/8/1800		
Charles Lucas . .	Ensign . . .	31/8/1791	Adjutant, 2nd Bn.	1799
	Lieutenant . .	6/8/1794	Invalided . .	26/6/1810
	Captain-Lieutenant	14/8/1800	Died at Vizagapatam	25/11/1816
	Captain . .	1/1/1803		
	Major . . .	21/10/1809		
George Pippard .	Lieutenant . .	6/8/1794	2nd Battalion	
Michael Egan . .	Lieutenant . .	14/7/1795	1st Battalion	

Names.	Appointments with Dates.		Remarks with Dates.	
Thomas Little	Ensign	30/12/1795	Adjutant, 1st Bn.	1799
	Lieutenant	8/1/1796	Died at Nattore	30/4/1815
	Captain	19/5/1804		
	Major	27/6/1810		
Henry Davy	Lieutenant	8/1/1796	1st Battalion	
Francis Evans	Lieutenant	4/3/1797	Adjutant, 2nd Bn.	27/8/1800
Joseph Knowles, C.B.	Ensign	12/3/1796	1st Battalion	
	Lieutenant	20/8/1797	Transferred	1819
	Captain-Lieutenant	1/5/1804		
	Captain	11/10/1804		
	Major	3/6/1812		
Charles R. Lester	Lieutenant	29/11/1797	2nd Battalion	
	Captain-Lieutenant	11/10/1804	Died	24/11/1812
	Captain	15/11/1804		
John Rendall	Lieutenant	2/8/1798	1st Battalion	
George Wilson	Lieutenant	12/9/1798	2nd Battalion	
	Captain-Lieutenant	15/11/1804	Died	17/3/1809
	Captain	15/11/1804		
Hastings M. Kelly	Lieutenant	24/9/1798	2nd Battalion	
			Transferred	17/6/1800
H. H. Pepper	Lieutenant	26/12/1798	1st Battalion	
	Captain-Lieutenant	...	Commandant, 1st Bn.	31/1/1821
	Captain	21/6/1807	,, 2nd Bn.	22/10/1822
			Transferred	1823
	Major	1/5/1815	Died in Pegu	25/7/1826
	Lieutenant-Colonel	31/1/1821		
John Carfrae	Ensign	9/9/1798	2nd Battalion	
	Lieutenant	26/12/1798	Commandant, 2nd Battalion	1/5/1824
			Retired	1/6/1824
	Captain-Lieutenant	21/6/1807		
	Captain	14/11/1807		
	Major	22/10/1818		
	Lieutenant-Colonel	1/5/1824		
Hugh Dalrymple	Lieutenant	31/12/1800	1st Battalion	
			Transferred	9/7/1801
Arthur Molesworth, 35th N.I.	Lieutenant	1798	Adjutant, 2nd Bn.	1800
			Transferred	1800
W. Walker, 2nd Bat. 1st Regt.	Lieutenant	5/4/1798		
Colin Campbell, 1st Regt.	Major	...	Offg. Commandant	1799
			Killed at Sultanpet	6/4/1799
Edward Collins	Colonel	3/10/1792	Colonel of 3rd Regt.	12/7/1796
			Transferred	1/1/1800
			Retired	28/2/1801
Barry Close	Lieutenant-Colonel	29/11/1797	Commandant, 1st Bn.	29/11/1797
John Chalmers	Lieutenant-Colonel	31/7/1799	Commandant, 2nd Battalion	31/7/1799
			Transferred	13/1/1801
John Bannerman	Major	29/9/1798	Commandant, 1st Battalion	17/6/1800
	Lieutenant-Colonel	17/6/1800	Transferred	29/6/1800
John Kennet	Ensign	16/7/1778	Commandant, 2nd Battalion	13/1/1801
	Lieutenant	12/3/1782	Died at Poonah	23/6/1803
	Captain	1/6/1796		
	Major	31/7/1799		
William Sheppard	Captain	8/1/1796	Commandant, 1st Bn.	1/5/1804
	Major	17/6/1800		
	Lieutenant-Colonel	1/5/1804		
Alexander Allan	Captain	8/1/1796	2nd Battalion	
	Major	23/6/1803	Retired	14/11/1804

Names.	Appointments with Dates.		Remarks with Dates.	
Poole H. Vesey	Captain	29/11/1797	1st Battalion	
	Major	1/5/1804	1st Battalion	
	Lieutenant-Colonel	6/12/1806	Commandant, 2nd Battalion	21/6/1807
			Retired	12/10/1808
Rouliston Robinett	Captain	12/9/1798	2nd Battalion	
			Died	14/8/1800
Charles Trotter	Captain	16/12/1798	1st Battalion	
			Transferred	11/10/1804
Charles Burton Phillipson	Captain	31/7/1799	2nd Battalion	
			Transferred	17/6/1800
Thomas Boles	Ensign	8/4/1788	2nd Battalion	
	Lieutenant	7/6/1792	1st Battalion	12/10/1808
	Captain	10/12/1799	Invalided	21/10/1809
James Welsh	Captain-Lieutenant	10/12/1799	1st Battalion	
	Captain	17/6/1800	Adjutant and Q.M.	1799
	Major	11/7/1806	2nd Battalion	12/10/1808
Patrick Joyes	Lieutenant	6/1/1793	Adjutant and Q.M., 1st Battalion	1798
Thomas Baynes	Lieutenant	18/3/1794	Adjutant and Q.M., 2nd Battalion	1798
John Lloyd Jones	Lieutenant	6/3/1793	2nd Battalion	
	Captain-Lieutenant	17/6/1800		
	Captain	17/6/1800		
Urban Vigors	Colonel	1/1/1800	Colonel of 3rd Regt.	1/1/1800
			Transferred	1/1/1803
William Kenny	Lieutenant-Colonel	1/1/1800	Commandant, 2nd Battalion	1/1/1800
			Transferred	13/1/1801
			Died	1/5/1804
James Innes	Lieutenant-Colonel	1/1/1800	Commandant, 1st Bn.	1/1/1800
			,, 2nd Bn.	5/3/1801
			Died	23/4/1804
James Graham	Major	1/1/1800	1st Battalion	
Francis Aiskill	Major	1/1/1800	2nd Battalion	
Samuel William Ogg	Captain	1/1/1800	1st Battalion	
	Major	7/1/1802	2nd Battalion	
	Lieutenant-Colonel	4/8/1816	Transferred in 12/12/1805, and reported in 1816	
Alexander Orr	Captain	1/1/1800	1st Battalion	
	Major	24/4/1804	2nd Battalion	
Thomas Currie	Captain	1/1/1800	1st Battalion	
			Transferred	1803
John Molloy	Captain	1/1/1800	2nd Battalion	
Charles Thomas Cooper	Captain	1/1/1800	2nd Battalion	
Galbraith Hamilton	Captain	17/6/1800	2nd Battalion	
	Major	21/9/1804		
George Lang	Captain-Lieutenant	17/6/1800	2nd Battalion	
	Captain	7/1/1802	1st Battalion	
	Major	27/1/1806		
Edward Gibbings	Lieutenant-Colonel	13/1/1801	Commandant, 2nd Battalion	13/1/1801
			Commandant, 1st Bn.	5/3/1801
Francis Aiskill	Major	1/1/1800	Commandant, 1st Bn.	1804
			Invalided	5/6/1804
Arthur Frith	Captain-Lieutenant	7/1/1802	1st Battalion	
	Captain	20/2/1802		
Thomas Sydenham	Lieutenant	1/1/1800	1st Battalion	
	Captain	26/3/1802		
Charles Hodgson	Lieutenant	1/1/1800	2nd Battalion	
	Captain	3/6/1802		
George Sinclair	Lieutenant	1/1/1800	Adjutant, 2nd Bn.	1/1/1800

Names.	Appointments with Dates.		Remarks with Dates.	
George Munro .	Lieutenant .	1/1/1800	1st Battalion	
	Captain-Lieutenant	10/11/1802		
Francis Innes .	Lieutenant .	1/1/1800	1st Battalion	
Robert E. Langford .	Lieutenant .	1/1/1800	1st Battalion	
Sydenham Smith .	Lieutenant .	1/1/1800	1st Battalion	
	Captain .	21/9/1804		
John M'Bean .	Lieutenant .	1/1/1800	2nd Battalion	
	Captain .	21/9/1804		
George Latten Lambert	Lieutenant .	1/1/1800	Adjutant, 1st Bn. .	8/5/1803
	Captain-Lieutenant	21/9/1804		
	Captain .	27/1/1806		
John William White	Lieutenant .	1/1/1800	Adjutant, 2nd Bn.	2/6/1802
	Captain-Lieutenant	27/1/1806		
William Moore .	Lieutenant .	1/1/1800	Adjutant, 1st Bn. .	14/9/1804
Richard West .	Lieutenant .	15/12/1800	2nd Battalion	
William Bennett .	Lieutenant .	15/12/1800	1st Battalion	
William Bruce .	Lieutenant .	15/12/1800	2nd Battalion	
			Resigned .	11/8/1804
Edward Norris .	Lieutenant .	15/12/1800	1st Battalion	
Henry Swayne .	Lieutenant .	15/12/1800	1st Battalion	
Edward Malton .	Lieutenant .	16/12/1800	2nd Battalion	
John Duncombe .	Ensign .	1/1/1800	2nd Battalion	
	Lieutenant .	20/7/1801		
Charles Cunningham	Ensign .	1/1/1800	2nd Battalion	
	Lieutenant .	20/7/1801		
Henry John Wilkinson	Ensign .	1/1/1800	Adjutant, 2nd Bn.	5/7/1805
	Lieutenant .	20/7/1801		
J. B. Scouler .	Ensign .	1/1/1800	2nd Battalion	
Robert Spears .	Ensign .	1/1/1800	1st Battalion	
Frederick Langley .	Ensign .	1/1/1800	1st Battalion	
	Lieutenant .	20/7/1801	Resigned .	5/6/1804
Henry Vincent .	Ensign .	1/1/1800	1st Battalion	
	Lieutenant .	20/7/1801		
John Beard .	Ensign .	1/1/1800	2nd Battalion	
Mark Cubbon .	Ensign .	1/1/1800	2nd Battalion	
Francis James .	Lieutenant .	1/1/1800	Adjutant, 1st Bn. .	1/1/1800
	Captain-Lieutenant	8/5/1803		
	Captain .	19/1/1804		
Michael Smith .	Lieutenant .	15/12/1800	Died .	7/6/1810
James Barter .	Surgeon .	17/10/1800	1st Battalion	
J. B. Greaves .	Ensign .	15/12/1800	1st Battalion	
Alexander Stewart .	Ensign .	15/12/1800	1st Battalion	
	Lieutenant .	...	Died .	10/6/1820
	Captain-Lieutenant	21/10/1809		
	Captain .	27/6/1810		
John Armstrong .	Lieutenant .	...	Died .	11/3/1804
John Barrett .	Lieutenant .	...	Died .	9/9/1806
James Walker .	Ensign .	15/12/1800	1st Battalion	
	Lieutenant .	1801		
	Captain-Lieutenant	1809		
	Captain .	18/3/1809		
	Major .	31/1/1821		
William Colhown .	Surgeon .	14/4/1801	2nd Battalion	
			Transferred .	1810
—— Jones .	Assistant-Surgeon .	1801	1st Battalion	
			Transferred .	19/8/1800
James Annesley .	Assistant-Surgeon .	1801	1st Battalion	
William Jones .	Lieutenant .	20/7/1801	1st Battalion	
Frederick Bruce .	Lieutenant .	20/7/1801	2nd Battalion	
James Bedds .	Lieutenant .	1801	Died .	4/7/1805
Thomas T. Stevenson	Lieutenant .	...		
	Captain .	21/10/1809	Adjutant, 2nd Bn.	14/10/1804
—— Capper .	Lieutenant .	1801	Died .	18/3/1809

Names.	Appointments with Dates.		Remarks with Dates.	
W. Plenderleath	Ensign	19/10/1801	1st Battalion	
George Hunter	Ensign	1801	Adjutant, 2nd Bn.	12/1/1808
	Lieutenant	20/7/1801	Commandant, 2nd Battalion	1/6/1824
	Captain-Lieutenant	18/3/1809		
	Captain	27/6/1810		
	Major	1/5/1824		
	Lieutenant-Colonel	5/9/1827		
P. G. Hill	Ensign	7/1/1802		
	Lieutenant	1802	Adjutant, 1st Bn.	1804
			Died	23/10/1806
—— Read	Major	...	Died	19/5/1804
Alexander Turner	Ensign	20/7/1801	1st Battalion	
	Lieutenant	17/12/1802		
	Captain-Lieutenant	3/6/1812		
	Captain	25/11/1812		
James Tagg	Lieutenant	20/7/1801		
	Captain	1/5/1815	Died at Kamptee	13/12/1825
John Wilson	Lieutenant	20/7/1801	Adjutant, 1st Bn.	23/10/1806
	Captain	1/5/1815	Commandant	25/5/1833
	Major	7/7/1827	Retired	30/5/1833
	Lieutenant-Colonel	25/5/1833		
John Baxter	Ensign	7/1/1802	2nd Battalion	
	Lieutenant	1803	Transferred	11/10/1804
Christopher Kingdom	Lieutenant	1/5/1802		
B. Humpage	Assistant-Surgeon	15/9/1802	2nd Battalion	
			Transferred	1802
Hans Gordon	Assistant-Surgeon	15/9/1802	1st Battalion	
W. Holloway	Assistant-Surgeon	15/9/1802	1st Battalion	
A. Inverarity	Assistant-Surgeon	15/9/1802	2nd Battalion	
			Transferred	24/1/1803
—— Varty	Lieutenant	1802	Transferred	11/10/1804
A. Babington	Surgeon	14/10/1802	1st Battalion	
Ambrose B. Perkins	Lieutenant	17/12/1802	1st Battalion	
			Died	25/4/1814
John Harlem	Assistant-Surgeon	24/1/1803	2nd Battalion	
			Transferred	
Leonard Cooper	Lieutenant	13/15/1803	1st Battalion	
David Binny	Lieutenant	3/7/1803	1st Battalion	
			Died	3/6/1807
Charles Watson Yates	Lieutenant	23/9/1803	2nd Battalion	
—— Haliburton	Lieutenant-Colonel	...	Commandant, 2nd Battalion	1804
			Transferred	19/5/1804
David Dundas Hamilton	Lieutenant	19/1/1804	1st Battalion	
Thomas W. Dawson	Lieutenant	11/3/1804	1st Battalion	
			Died	19/5/1812
Henry Jonathan Cope	Lieutenant	24/4/1804	2nd Battalion	
William Kinsey	Colonel	1/5/1804	Transferred	1825
			Died	6/4/1837
A. Sweedland	Lieutenant	1/5/1804	2nd Battalion	
John Cuppage	Lieutenant-Colonel	30/5/1804	Commandant, 2nd Battalion	30/5/1804
			Transferred	31/5/1804
R. Warne	Lieutenant-Colonel	31/5/1804	Commandant, 2nd Battalion	31/5/1804
			Transferred	1805
Allan Macleod	Lieutenant	6/6/1804	2nd Battalion	
Alexander Robertson	Lieutenant-Colonel	9/8/1804	Commandant, 1st Battalion	
Stephen Rolleston	Lieutenant	21/9/1804	Invalided	30/11/1810
William Isaacke	Lieutenant	21/9/1804	1st Battalion	

Names.	Appointments with Dates.		Remarks with Dates.	
Josiah Stewart	Lieutenant	21/9/1804	1st Battalion	
John Stewart	Lieutenant	21/9/1804	2nd Battalion	
William Kelso	Lieutenant	21/9/1804	2nd Battalion	
Arthur Bentley	Lieutenant	21/9/1804	2nd Battalion	
Ralph Gore	Lieutenant	21/9/1804	2nd Battalion	
			Died	30/1/1813
Fraser M. Goble	Lieutenant	21/9/1804	2nd Battalion	
			Died	14/2/1813
Henry Dumas	Lieutenant	21/9/1804	Cashiered	10/2/1813
J. L. Hooffstetter	Lieutenant	21/9/1804		
	Captain-Lieutenant	4/4/1816		
Robert Montague Strange, 10th Regt.	Lieutenant-Colonel	28/3/1805	Commandant, 1st Battalion	28/3/1805
David Agnew	Lieutenant	17/7/1805	2nd Battalion	
	Captain-Lieutenant	1/10/1816		
	Captain	1/10/1816		
Ambrose Herne Colberg	Lieutenant	17/7/1805	2nd Battalion	
	Captain	1/10/1818	Died at Chunapatam	15/10/1827
William H. Jeffrey	Lieutenant	17/7/1805	2nd Battalion	
			Died	19/11/1809
Hastings Archibald Moncrieff	Lieutenant	17/7/1805	1st Battalion	
John St. George Ferns	Lieutenant	17/7/1805	1st Battalion	
Nathaniel Alves	Lieutenant	20/9/1805	1st Battalion	
—— Lindsay, 7th Regt.	Lieutenant-Colonel	12/12/1805	Commandant, 2nd Battalion	12/12/1805
			Transferred	13/12/1805
Alexander Baillie, 9th Regt.	Lieutenant-Colonel	13/12/1805	Commandant, 2nd Battalion	13/12/1805
			Transferred	1806
Robert John Marr	Ensign	27/6/1806	2nd Battalion	
	Lieutenant	14/11/1809		
	Captain-Lieutenant	27/6/1810		
	Captain	6/12/1821		
John Herring	Ensign	27/6/1806	2nd Battalion	
	Lieutenant	20/11/1809		
Charles Brumfield	Ensign	27/6/1806	1st Battalion	
John Ross	Ensign	27/6/1806	2nd Battalion	
John Gwynne	Ensign	27/6/1806	1st Battalion	
Henry Harkness	Ensign	27/6/1806	2nd Battalion	
Richard Bullevant	Ensign	27/6/1806	1st Battalion	
William Henderson	Ensign	27/6/1806	2nd Battalion	
—— M'Kerras	Lieutenant-Colonel	1806	Commandant, 2nd Battalion	
			Died	11/7/1806
C. B. Robinson	Lieutenant	9/9/1806	2nd Battalion	
	Captain	22/10/1818	Died at Madras	18/6/1824
John Fyfe	Lieutenant	22/10/1806	1st and 2nd Bns.	
	Captain	10/6/1820		
Samuel Revière	Assistant-Surgeon	15/2/1807	1st Battalion	
William Williamson	Lieutenant	22/5/1807	Adjutant, 1st Bn.	5/10/1815
	Captain	31/1/1821		
Robert Inverarity	Lieutenant	24/6/1807	Adjutant, 2nd Bn.	19/2/1813
	Brevet Captain	23/5/1821	Quartermaster and Interpreter, 2nd Battalion	4/6/1819
			Retired in England	14/6/1824
William Shepperd	Ensign	3/7/1807		
	Lieutenant	27/6/1810	Died	17/1/1813
Bridgewater Meredith	Ensign	3/7/1807	Died in Camp at Ashta	22/11/1817
	Lieutenant	27/6/1810		

Names.	Appointments with Dates.		Remarks with Dates.	
Thomas Robinson	Ensign	3/7/1807	1st Battalion	
	Lieutenant	1/12/1810		
Henry Dixon	Ensign	3/7/1807		
	Lieutenant	19/2/1812		
William Miles Blake	Ensign	3/7/1807	2nd Battalion	
J. Hector Greig	Ensign	3/7/1807	1st Battalion	
	Lieutenant	23/3/1812		
Edward Rule	Ensign	3/7/1807	2nd Battalion	
	Lieutenant	24/7/1815		
—— Innes, 22nd Regt.	Lieutenant-Colonel	16/7/1807	Commandant, 1st Battalion	16/7/1807
John Pierson	Ensign	7/10/1807	2nd Battalion	
	Lieutenant	14/11/1807	Died	18/2/1812
Gilbert W. Eccles	Ensign	7/10/1807	1st Battalion	
	Lieutenant	21/10/1809	Died	22/4/1812
Thomas Sutton	Assistant-Surgeon	3/2/1808	2nd Battalion	
			Transferred	3/10/1808
J. Long	Lieutenant-Colonel	...	Commandant, 2nd Battalion	1808
			Retired	1/10/1808
Stephen Purrock	Assistant-Surgeon	3/10/1808	1st Battalion	
W. S. Anderson	Assistant-Surgeon	3/10/1808	2nd Battalion	
			Transferred	16/11/1812
—— Corner, 20th Regt.	Lieutenant-Colonel	12/10/1808	Commandant, 2nd Battalion	12/10/1808
George Martin	Lieutenant-Colonel	2/1/1809	Commandant, 2nd Battalion	2/1/1809
			Died at Bellary	17/6/1815
—— Cadenskie, H.M.'s 80th Regt.	Lieutenant	9/8/1809	Acting - Adjutant, 2nd Battalion	9/8/1809
James Brunton	Lieutenant-Colonel	8/12/1807	Commandant, 1st Battalion	1809
			Died in England	6/11/1810
Robert Barclay	Lieutenant-Colonel	15/10/1809	Commandant, 2nd Battalion	15/10/1809
			Transferred	9/2/1810
William Orrok, 19th Regt.	Lieutenant-Colonel	9/2/1810	Commandant, 2nd Battalion	9/2/1810
			Died at Seringapatam	26/6/1810
Edward Williams	Ensign	24/10/1809		
	Lieutenant	3/6/1812		
John Gordon Rorison	Ensign	27/1/1810		
	Lieutenant	27/12/1812	Transferred	3/8/1838
	Captain	1/5/1824		
	Major	31/5/1833		
Henry Webber	Lieutenant-Colonel	28/2/1810	Commandant, 2nd Battalion	1811
			Transferred	1813
			Died	21/8/1833
William Roper	Ensign	5/4/1810		
	Lieutenant	17/1/1813		
David Armstrong	Ensign	6/4/1810		
	Lieutenant	17/1/1813		
W. Raynsford Taylor	Ensign	5/5/1810		
	Lieutenant	31/1/1813	Transferred to Civil	14/9/1813
Joseph Leggatt	Ensign	18/5/1810		
	Lieutenant	11/2/1813		
George Walker	Ensign	27/6/1810		
	Lieutenant	15/2/1813		
John Grant	Surgeon	1810	2nd Battalion	
			Transferred	8/1/1811
A. Tozer	Surgeon	8/1/1811	1st Battalion	

Names.	Appointments with Dates.		Remarks with Dates.	
Henry Nash, 6th Regt.	Lieutenant-Colonel	9/10/1811	Commandant, 1st Battalion	9/10/1811
Lewis Crowther, 22nd Regt.	Ensign . . .	26/6/1812	2nd Battalion	
	Lieutenant . .	15/2/1813	Invalided . .	13/9/1815
George Dods, M.V. Battalion	Ensign . . .	26/6/1812	Adjutant, 1st Bn. .	29/3/1819
	Lieutenant . .	10/10/1813	Adjutant, 2nd Bn.	4/6/1824
	Captain . .	15/6/1824	Commandant .	23/12/1842
	Major . . .	3/8/1838	Retired . .	15/3/1845
	Lieutenant-Colonel	23/12/1842		
John Stodart, 17th Regt.	Ensign . . .	26/6/1812	1st Battalion	
	Lieutenant . .	26/4/1814		
A. Campbell, 5th Regt.	Assistant-Surgeon .	16/11/1812	2nd Battalion	
			Transferred . .	3/9/1816
Richard Budd . .	Ensign . . .	11/6/1812	1st Battalion	
			Transferred . .	20/3/1813
John Jones, 19th Regt.	Ensign . . .	20/3/1813	1st Battalion	
	Lieutenant . .	26/4/1814	Died at Vellore .	18/3/1821
William J. D. Glen .	Ensign . . .	13/7/1813	1st Battalion	
	Lieutenant . .	4/4/1816	Killed at Mahed-pore	21/12/1817
Thomas Clemons .	Ensign . . .	6/7/1813	Transferred . .	1818
James Maxtone	Ensign . . .	6/7/1813	Adjutant, 2nd Bn.	31/5/1819
	Lieutenant . .	1/10/1816	Died at Palamcotta	21/10/1828
	Captain . .	23/4/1825		
M. Wilkes . .	Lieutenant-Colonel	1813	Commandant, 2nd Battalion	1813
			Transferred . .	4/8/1816
William Blackburne .	Lieutenant-Colonel	8/1/1814	Commandant, 1st Bn.	8/1/1814
Edwin J. Johnson .	Lieutenant . .	1/5/1815		
James Briggs . .	Ensign . . .	11/7/1815	Q.M. Int. & P. M.	4/6/1824
	Lieutenant . .	27/10/1817	Retired . .	24/12/1842
	Captain . .	7/7/1827		
	Major . . .	?		
J. Macaulay, 2nd Bn. 4th Regt.	Surgeon . .	21/8/1815		
James Robins . .	Lieutenant . .	12/10/1815	Transferred . .	22/6/1825
Kenneth Macaulay .	Surgeon . .	26/10/1815	Transferred . .	3/9/1816
—— Harwood . .	Assistant-Surgeon .	1815		
M. M'Intosh . .	Surgeon . .	1815	Transferred . .	21/8/1815
Samuel William Ogg, 1st Bn. 19th Regt.	Lieutenant-Colonel	4/8/1816	Transferred . .	18/8/1819
John Cormick . .	Surgeon . .	3/9/1816		
Adam Stevenson, Rifle Corps	Assistant-Surgeon .	3/9/1816	Transferred . .	25/9/1820
C. Budd . . .	Lieutenant . .	3/12/1817		
Thomas R. Barton .	Lieutenant . .	4/6/1818		
Thomas Dallas .	Lieutenant . .	4/6/1818	Q.M. Int. & P.M., 2nd Bn.	9/3/1822
Edward Rogers .	Lieutenant . .	4/6/1818	Died . . .	3/2/1829
	Captain . .	16/10/1827		
William Munro .	Lieutenant-Colonel	9/6/1818		
Henry Bowen . .	Lieutenant-Colonel	18/8/1818		
Percy L. Harvey .	Ensign . . .	1818		
	Lieutenant . .	12/8/1818		
Thomas A. Fraser .	Lieutenant-Colonel	28/1/1819		
C. Elliot . . .	Lieutenant . .	19/5/1819		
Charles Fladgate .	Lieutenant . .	13/6/1819	Q.M. Int. & P.M.	1/7/1825
	Captain . .	13/2/1829		
	Major . . .	24/12/1842		
Charles Richardson .	Lieutenant . .	13/6/1819	Died . . .	28/12/1823
Walter L. Williams .	Lieutenant . .	13/6/1819		
John Deane . .	Lieutenant . .	13/6/1819		
William Macqueen .	Lieutenant . .	13/6/1819		

Names.	Appointments with Dates.		Remarks with Dates.	
Rich. J. Charlton	Lieutenant	13/6/1819		
James Fraser	Ensign	3/7/1819		
C. R. Bradstreet	Ensign	1819		
J. M. Boyes	Ensign	1819		
Fred Smith	Ensign	1819	Transferred	13/7/1819
T. G. E. G. Kenny	Ensign	6/3/1820	Adjutant	27/5/1825
	Lieutenant	14/7/1823		
James Power	Ensign	6/3/1820		
	Lieutenant	1/7/1823		
John P. Parsons	Lieutenant	7/3/1820	Died at Cuddalore	13/6/1823
John C. Glover	Lieutenant	7/3/1820	Transferred	11/3/1845
	Captain	12/11/1830		
C. S. Buxton	Lieutenant	18/6/1820		
Thomas Edwards	Assistant-Surgeon	25/9/1820	2nd Battalion	
	Surgeon	4/6/1824	Transferred	5/6/1827
Chris. Dennett	Ensign	1820		
Chas. H. Graham	Ensign	1820		
Dering Addison	Ensign	1820		
	Lieutenant	13/6/1819		
John Sheil	Lieutenant	2/1/1821	...	1845
	Captain	25/3/1833		
	Major	11/3/1845		
J. Mackenzie, C. B., 25th Regt.	Lieutenant-Colonel	6/2/1821	Commandant	6/2/1821
			Transferred	22/10/1822
James F. Leslie	Ensign	13/2/1821	Died at Singapore	31/5/1842
	Lieutenant	1/5/1824		
	Captain	26/3/1842		
John M. George	Ensign	13/2/1821		
William H. Short	Ensign	13/2/1821		
	Lieutenant	28/12/1823		
Alexander Campbell	Ensign	13/2/1821	Died	1823
George W. Watson	Ensign	13/2/1821		
	Lieutenant	15/6/1824		
John Johnstone	Ensign	13/2/1821		
Joseph Standiver Sherman	Lieutenant	14/10/1821	1st Battalion	
	Captain	3/8/1838	Died at Ellore	26/3/1842
Robert Davidson	Assistant-Surgeon	1821		
John Everest	Ensign	27/3/1822	Died at Madras	7/3/1832
	Lieutenant	22/6/1825		
Thomas J. Adams	9th Ensign	3/2/1823		
Horace C. Beevor	Ensign	2/5/1823	Q.M., Int., & P.M.	3/3/1829
	Lieutenant	7/7/1827	Transferred	8/1/1846
	Captain	31/5/1842		
C. T. G. Bishop	Lieutenant-Colonel	1823	Commandant	1823
			Retired	1825
—— Richardson	Lieutenant	1823		
H. F. Smith, C.B.	Lieutenant-Colonel	1823	Commandant	1823
			Transferred	21/3/1824
A. Limond	Lieutenant-Colonel	21/3/1824	Commandant	21/3/1824
			Transferred	3/6/1824
William M'Leod	Col.-Commandant	3/6/1824	Retired	16/11/1838
Fraser P. Stewart	Lieutenant-Colonel	3/6/1824	Commandant	3/6/1824
			Retired	23/1/1826
George C. Hughes	4th Ensign	5/7/1824	Commandant	2/11/1854
	Lieutenant	16/10/1827	Retired	1857
	Captain	23/12/1842		
	Major	6/8/1846		
	Lieutenant-Colonel	2/11/1854		
James Wyse	Surgeon	1824		
W. Elliot Lockhart	Ensign	6/5/1825	Transferred	25/9/1826
J. J. G. Congdon	4th Ensign	23/6/1825	Transferred	21/7/1825

Names.	Appointments with Dates.		Remarks with Dates.	
James William Gammell Kenny	Ensign . . .	8/1/1826		
	Lieutenant . .	26/5/1833		
	Captain . .	24/1/1845		
	Major . . .	14/2/1855		
	Lieutenant-Colonel	4/2/1861		
Fidelio Robert Trewman	Ensign . . .	8/1/1826		
	Lieutenant . .	3/2/1829		
Frederick Bowes .	Lieutenant-Colonel	23/1/1826	Commandant .	23/1/1826
			Retired . .	1828
G. A. Kerklots, M.D., 10th N.I.	Surgeon . .	12/3/1826	Died at Wallajabad	8/1/1834
Henry Richard Dardis	5th Ensign . .	23/8/1826	Died at Gooty .	15/5/1827
J. E. Hughes, 47th N.I.	Ensign . . .	6/9/1826		
—— Gibson . .	Captain . .	1826		
Charles Gordon, 6th N.I.	4th Ensign . .	12/2/1827	Retired . .	14/2/1855
	Lieutenant . .	8/3/1832		
	Captain . .	24/12/1842		
	Major . . .	16/11/1854		
Richard Ramsden, 18th N.I.	5th Ensign . .	12/2/1827		
Josiah Smith, 9th N.I.	4th Ensign . .	24/9/1827	Retired . .	2/10/1855
	Lieutenant . .	3/8/1838		
	Captain . .	7/3/1845		
Edward Slack, 21st N.I.	5th Ensign . .	24/9/1827	Q.M., Int., & P.M.	22/3/1833
	Lieutenant . .	15/9/1839	Transferred . .	21/9/1840
—— Cadell . .	Lieutenant-Colonel	...	Transferred . .	5/9/1827
B. B. Parlby, C.B., 19th N.I.	Lieutenant-Colonel	23/1/1828	Transferred . .	31/12/1828
James Ross Arrow .	Ensign . . .	27/5/1828	Transferred . .	25/8/1828
J. Underwood, 3rd Battalion Artillery	Surgeon . .	21/9/1828	Transferred . .	20/10/1834
Thomas King, 44th N.I.	Lieutenant-Colonel	31/12/1828	Commandant .	31/12/1828
			Transferred . .	29/7/1831
Frederick C. Hawkins	Ensign . . .	11/2/1829		
Edward T. Cox .	Ensign . . .	20/8/1829	Transferred . .	1835
John O'Neil . .	Assistant-Surgeon .	17/10/1829	Transferred . .	15/12/1834
W. Milne, 29th N.I.	Lieutenant-Colonel	29/7/1831	Commandant .	29/7/1831
			Retired . .	25/5/1833
Josiah Stewart . .	Lieutenant-Colonel	4/2/1832	Commandant .	4/2/1832
			Retired . .	1834
G. S. Mardall . .	Ensign . . .	24/9/1833		
David Blair . .	Ensign . . .	24/9/1833	Transferred . .	27/10/1834
Edward Cadogan, 33rd N.I.	Lieutenant-Colonel	30/5/1833		
Henry Corbett Taylor	Ensign . . .	16/6/1834		
A. N. Magrath .	Surgeon . .	20/10/1834	Transferred . .	15/12/1834
—— Templer . .	Ensign . . .	1834	Transferred . .	27/10/1834
Robert Crowe . .	Ensign . . .	1834	Transferred . .	27/10/1834
Walter Frederick Goodwyn, 19th N.I.	3rd Ensign . .	27/10/1834	Q.M. and Int. . .	14/1/1840
	Lieutenant . .	8/10/1839	Transferred . .	7/3/1848
	Captain . .	11/3/1845		
Arthur Robinson, 16th N.I.	Ensign . . .	13/11/1834	Adjutant . .	24/12/1839
	Lieutenant . .	21/9/1840		
	Captain . .	3/1/1846		
Kenneth Macaulay, 20th N.I.	Surgeon . .	15/12/1834	Transferred . .	5/9/1835
H. Goodall, 20th N.I.	Assistant-Surgeon .	15/12/1834	Transferred . .	20/3/1840
Thomas Hayes Brown Ludlow	Lieutenant . .	1834		27/10/1834
J. Dalmahoy . .	Surgeon . .	5/9/1835	Transferred . .	20/6/1836
Herbert Main Dobbie	Ensign . . .	21/7/1835	Transferred . .	14/8/1835
Rodney Chas. Kempt	Ensign . . .	21/7/1835	Transferred . .	14/8/1835
C. F. Gordon . .	Ensign . . .	1835	Transferred . .	14/8/1835

Names.	Appointments with Dates.		Remarks with Dates.	
E. H. L. Moore	Ensign	1835	Transferred	14/8/1835
Samuel Higginson	Surgeon	20/6/1836	Died at Maulmain	12/9/1838
—— Lawrie	Major	1838		
—— Wheeler	Lieutenant	...	Adjutant	1838
I. T. Maule	Surgeon	19/1/1839	Transferred	17/3/1842
George Paxton	Ensign	26/6/1839		
F. J. M. Mason	Ensign	26/6/1839	Transferred	15/9/1839
Charles Woodland	Ensign	26/6/1839		
C. E. M. Walker	Ensign	26/6/1839		
Hon. P. O. Murray	Ensign	26/6/1839		
S. G. Prendergast	Ensign	3/7/1839	...	1847
	Lieutenant	3/10/1840		
	Captain	6/8/1846		
W. H. West	Ensign	12/9/1839	Transferred	5/12/1839
C. J. Bradley	Ensign	12/9/1839	Transferred	5/12/1839
W. A. Greenlaw	Ensign	21/9/1839		
Edward Beecher Marsack, 33rd N.I.	3rd Ensign	5/12/1839	Wing commander	25/3/1877
	Lieutenant	26/3/1842	Commandant (Offg.)	1865
			Commandant	29/6/1877
	Captain	15/12/1846	Retired	12/8/1878
	Major	20/9/1860		
	Lieutenant-Colonel	1/11/1865		
	Colonel	1876		
Edmund Francis Burton	4th Ensign	5/12/1839	Q.M. and Int.	3/6/1845
	Lieutenant	31/5/1842	Adjutant	7/7/1846
	Captain	7/3/1848	Commandant	1/11/1865
	Major	23/8/1861	Retired	15/3/1866
	Lieutenant-Colonel	1/11/1865		
S. W. Steel, C. B., 32nd N.I.	Lieutenant-Colonel	27/1/1840	Commandant	27/1/1840
			Retired	1843
Arnold Ward	Ensign	13/6/1840	Transferred	22/7/1840
W. C. Brackenbury	Ensign	6/8/1840	Transferred	1840
M. Riddell, 15th N.I.	3rd Ensign	28/12/1840	Transferred	1/2/1853
	Lieutenant	16/7/1842		
	Captain	17/3/1852		
Francis Albany Brooking	4th Ensign	28/12/1840	Adjutant	24/11/1848
	Lieutenant	23/12/1842		
	Captain	1/2/1853		
	Major	16/7/1864		
A. M. Cleghorn	Ensign	1839	Transferred	5/12/1839
Henry Tucker Campbell	Ensign	1841		
	Lieutenant	7/2/1845		
	Captain	14/2/1855		
	Major	1/10/1867		
Charles William Taylor	Ensign	28/5/1841	Adjutant	21/2/1851
	Lieutenant	24/12/1842	2nd in command & W.C.	1/11/1865
	Captain	16/11/1854		
	Major	29/5/1865	Retired	13/9/1870
	Lieutenant-Colonel	11/1867		
Robert Balfour	Lieutenant	1841		
B. Williams, 14th N.I.	Surgeon	17/3/1842	Transferred	25/8/1845
A. C. B. Neil, M.D.	Assistant-Surgeon	19/3/1842	Transferred	28/1/1843
Henry Rhodes Morgan	Ensign	6/8/1842		
	Lieutenant	7/3/1845		
	Captain	2/10/1855		
Basil Walter Marriott	5th Ensign	5/10/1842	Transferred	17/8/1851
	Lieutenant	11/3/1845		
Robert Ormsby Cary	Ensign	28/12/1842		
	Lieutenant	8/1/1846		
	Captain	23/11/1856		
P. F. Nicholson	Ensign	31/12/1842	Transferred	1844
J. Dorward, 39th N.I.	Assistant-Surgeon	28/1/1843	Transferred	8/11/1844

Names.	Appointments with Dates.		Remarks with Dates.	
J. A. W. F. Wilson	Ensign	15/6/1843	...	15/4/1853
	Lieutenant	6/8/1846		
Col. (Lt.-Gen.) W. M. C. Fraser	Colonel	1843	Transferred	9/3/1848
J. Pringle, M.D.	Assistant-Surgeon	8/11/1844	Transferred	16/9/1845
W. Burrell, 41st N.I.	Surgeon	9/1/1845	Transferred	29/5/1845
J. H. Winbolt, 22nd N.I.	Lieutenant-Colonel	15/3/1845	Commandant	15/3/1845
	Lieutenant-Colonel Commandant	9/3/1848	Retired	9/3/1848
H. A. Leveson	3rd Ensign	2/4/1845	...	30/9/1853
	Lieutenant	15/12/1846		
William Boardman	Ensign	26/5/1845	Officer Junior W.C.	25/10/1865
	Lieutenant	7/3/1848	Doing Duty Officer	1/11/1865
	Captain	29/10/1860		
A. Mackintosh, M.D.	Assistant-Surgeon	16/9/1845	Transferred	2/3/1847
G. A. C. Bright, 43rd N.I.	Surgeon	25/8/1845		
W. C. Hutton, 29th N.I.	Ensign	2/10/1845		
James Hale Warden	5th Ensign	15/6/1846	Q.M. and Int.	7/3/1848
	Lieutenant	17/8/1851		29/8/1861
	Captain	23/8/1861		
H. C. Roberts	4th Ensign	14/12/1846		
D. T. Morton	Assistant-Surgeon	2/3/1847	Transferred	27/11/1847
William Masterman Williams	4th Ensign	6/5/1847	Adjutant	21/11/1854
	Lieutenant	17/3/1853		
	Captain	2/3/1863		
Richard Kirwan Macquoid	5th Ensign	6/5/1847	Transferred	1864
	Lieutenant	1/2/1853		
	Captain	16/7/1864		
A. H. Ashley	Assistant-Surgeon	27/11/1847	Transferred	17/5/1848
R. Thorpe	Lieutenant-Colonel	9/3/1848	Transferred	4/10/1848
J. Cadenhead, 33rd N.I.	Assistant-Surgeon	17/5/1848		
Dashwood Charles Gordon Strettel	Ensign	13/8/1848		
	Lieutenant	30/9/1853		
George Hutton, 52nd N.I.	Lieutenant-Colonel	4/10/1848	Retired	3/3/1850
Barrington Frederick Heysham	Ensign	8/11/1848	Q.M. and Int.	27/4/1857
	Lieutenant	5/3/1853	Transferred	25/11/1867
	Captain	29/5/1865		
L. W. Watson	Lieutenant-Colonel Commandant	16/12/1848	Transferred	1859
J. W. Sherman, 31st N.I.	Surgeon	19/2/1849	Transferred	5/10/1849
W. Gilchrist	Surgeon	1849	Transferred	19/2/1849
S. T. Lyell	Assistant-Surgeon	19/2/1849	Transferred	12/5/1851
J. D. Stokes, 48th N.I.	Lieutenant-Colonel	3/3/1850	Transferred	1851
H. R. Oswald, M.D., 2nd Bn. Art.	Assistant-Surgeon	12/5/1851	Transferred	28/2/1852
Samuel Geo. Fred. Cooper, 1st M. Fus.	Ensign	20/9/1851	Transferred	11/1/1861
	Lieutenant	16/11/1854		
C. A. Browne	Lieutenant-Colonel	1851	Retired	2/11/1854
W. T. Groom	Ensign	1851	Transferred	20/9/1851
J. Brett, M.D., 35th N.I.	Assistant-Surgeon	28/2/1852		
Henry Hamilton Hooper, 28th N.I.	5th Ensign	22/3/1852		
	Lieutenant	14/2/1855		
A. Burlton Irving	5th Ensign	2/4/1853		
	Lieutenant	2/10/1855		
H. D. Hill, 44th N.I.	5th Ensign	13/8/1853	Transferred	25/8/1853

Names.	Appointments with Dates.		Remarks with Dates.	
C. A. Benson	4th Ensign	9/1/1854		
Malcolm M'Neil, 12th N.I.	Ensign	15/2/1854		
	Lieutenant	23/11/1856		
N. D. Robertson, 38th N.I.	5th Ensign	30/9/1854	Transferred	1856
W. Williamson	Assistant-Surgeon	3/3/1855	Transferred	7/11/1855
C. M. Moberly, 2nd E.L.I.	3rd Ensign	22/5/1855		
H. C. Wright	3rd Ensign	1/6/1855		
F. D. Plowden	4th Ensign	7/8/1855		
J. M'D. Smith	3rd Ensign	7/8/1855		
J. M'Donald	Assistant-Surgeon	7/11/1855	Transferred	16/7/1860
Wigram Arthur Cheek	3rd Ensign	20/11/1855	Adjutant	1/11/1865
	Lieutenant	23/11/1856		
	Captain	1/10/1867		
C. A. Liardet	4th Ensign	7/12/1855		
Joseph George Ellis Cameron, 6th N.I.	3rd Ensign	26/2/1856		
	Lieutenant	8/1/1858		
A. R. Oakes	4th Ensign	26/2/1856		
Richard Samuel Roberts, 12th N.I.	4th Ensign	6/5/1856	2nd Wing subaltern	29/6/1867
	Lieutenant	29/10/1860	Quartermaster	21/12/1867
	Captain	...	1st Wing subaltern	30/6/1875
			Transferred	1875
Daniel Hall Stevenson	Lt.-Col. (Bt.-Col.)	12/2/1857	Commandant	12/2/1857
			Transferred	22/10/1859
A. Chrystie, 17th N.I.	Ensign	21/4/1857		
Edward George Vere Holloway, 1st M. Fus.	3rd Ensign	21/4/1857	Transferred	1/6/1858
T. A. Duke	Lt.-Col. (Bt.-Col.)	1857	Commandant	...
			Transferred	12/6/1857
E. M. Stevenson, 45th N.I.	Attached as Ensign	4/1/1858	Transferred	1/3/1858
T. R. Tabuteau, 1st M. Fus.	3rd Ensign	16/2/1858	Transferred	6/9/1858
			Transferred	1/11/1865
John Briggs	Colonel	16/3/1858	Transferred	1/11/1859
R. T. Chapman	4th Ensign	23/8/1859		
J. W. Goldsworthy, 31st L.I.	Lt.-Col. (Bt.-Col.)	22/10/1859	Commandant	22/10/1859
			Transferred	21/1/1861
—— M'Leod	Lieutenant-Colonel	1859	Commandant	1859
			Transferred	22/10/1859
F. M. Mathias	Attached as Ensign	12/4/1860	Transferred	1860
R. E. Cox	Attached as Ensign	28/4/1860	Transferred	1860
H. E. Bustead	Assistant-Surgeon	16/7/1860	Transferred	15/3/1863
J. T. Clarke, 34th L.I.	Attached as Captain	1/12/1860	Transferred	20/12/1861
J. Hill, 41st N.I.	Lt.-Col. (Bt.-Col.)	21/1/1861	Commandant	21/1/1861
			Transferred	4/2/1861
T. H. Stotton	Lieutenant	4/3/1861	Transferred	1/7/1861
J. P. James	Lieutenant	23/8/1861	Quartermaster	22/2/1869
H. A. Hare	Captain	26/11/1861	Q.M. and Int.	26/11/1861
T. D. T. Dyer	Lieutenant-Colonel	8/1/1862	Commandant	8/1/1862
			Transferred	8/1/1862
R. A. Chadwick, 5th N.I.	Lieutenant	1862	Transferred	27/11/1862
—— Nelson	Lieutenant	1862	...	1862
A. H. H. Harvey	Lieutenant-Colonel	8/1/1862	Transferred	24/9/1862
E. J. Gunthorpe	Ensign	1862	Transferred	9/1/1862
A. C. Williams	Lieutenant	1858	Transferred	24/9/1864
E. G. Blenkinsop	Lieutenant	1858	Transferred	5/10/1864
J. A. Cox, M.D., 5th N.I.	Assistant-Surgeon	15/3/1863	Transferred	27/8/1863
T. G. Howell, I.M.S.	Assistant-Surgeon	1863	Transferred	14/12/1863

Names.	Appointments with Dates.		Remarks with Dates.	
L. W. Stewart, 11th N.I.	Assistant-Surgeon	2/11/1863	Transferred . .	2/5/1866
—— Beaman, I.M.S.	Assistant-Surgeon	14/12/1863	Transferred . .	24/8/1864
E. J. Watson, 23rd L.I.	Lieutenant . .	...		
	Adjutant . .	28/12/1863	Transferred . .	1864
W. Hands, Staff Corps	Captain . .	...		
	Ag. Q.M. and Interpreter	28/12/1863	Transferred . .	15/2/1864
C. B. Smith, 40th N.I.	Lieutenant . .	...		
	Ag. Q.M. and Interpreter	11/7/1864	Transferred . .	2/2/1865
W. N. Wroughton, 40th N.I.	Lieutenant . .	...		
	Offg. Adjutant .	26/10/1864	Transferred . .	1866
Major C. Pulley, 37th N.I.	Offg. Commandant	25/10/1865	Transferred . .	1867
Capt. (Brevet-Major) J. W. Rutherford, 47th N.I.	Wing Officer	1/11/1865	Transferred . .	27/10/1866
	Junior Wing Commandant	25/10/1865		
Lieut.-Col. W. T. Money, 7th N.I.	Commandant .	15/3/1866	Transferred . .	1/3/1867
Lieut. R. C. Sherrard, 44th Foot (Capt.)	Wing Officer .	7/7/1866	Died at Jubbulpur.	1888
Major A. G. W. Burn, 14th N.I.	Offg. Wing Officer	27/10/1866	Transferred . .	8/6/1874
	Wing Officer .	13/5/1867		
Lieut. H. M. Wratislaw, 14th N.I.	Attached . .	15/11/1866	Transferred . .	1867
Major Hailes . .	...	1867	Transferred . .	1867
Assist.-Surg. H. J. Beach, I.M.S.	In Medical Charge	19/1/1867	Transferred . .	3/5/1879
Ensign E. Moore (Lieut., Captain, Major, Lieut.-Col.)	2nd Wing Subaltern	8/2/1867	Retired . .	1888
	1st Wing Subaltern	8/2/1867		
	Adjutant . .	17/6/1867		
Captain G. E. H. Beauchamp, 45th N.I.	Quartermaster .	22/11/1865	Transferred . .	4/7/1870
	1st Wing Subaltern	17/6/1867		
	Offg. Wing Officer	23/11/1868		
Major H. T. Campbell; 28th N.I.	Wing Officer .	1/4/1867	Transferred . .	13/5/1867
Lieut.-Col. C. P. Y. Triscott, 25th N.I.	Commandant .	1/4/1867	Transferred . .	9/10/1871
Major C. M. White .	Junior Wing Commandant	9/1868	Transferred . .	29/4/1869
Capt. H. Coningham, 12th N.I.	Offg. Adjutant .	30/7/1869	Transferred . .	13/10/1869
Capt. H. H. Warrington (Major, Lieut.-Col.)	1st Wing Subaltern	4/7/1870	Retired . .	1886
Col. A. K. C. Kennedy, 17th N.I.	Offg. Commandant	13/9/1870	Transferred . .	27/7/1874
	Commandant .	9/10/1871		
Surgeon R. V. Power, I.M.S.	In Medical Charge	20/9/1870	Transferred . .	5/1/1878
Lieut.-Col. G. T. Hilliard, General Duty	W. O. and 2nd in command	13/10/1870	Transferred . .	10/11/1876
	Offg. Commandant	10/1/1875		
	Commandant .	8/1/1876		
Capt. H. G. Symons.	1st Wing Subaltern	1/12/1873		
Capt. C. L. Raikes, Staff Corps	Offg. 1st Wing Subaltern	26/3/1874		
	1st Wing Officer .	19/7/1874		
Lieut. H. I. Bett, 17th N.I.	Wing Officer .	8/6/1874	Transferred . .	31/12/1874
Capt. L. B. Byass, Staff Corps	Offg. 2nd Wing Officer	2/7/1874	Transferred . .	20/8/1875
Lt.-Col. R. C. Lavie	Offg. Wing Officer	1875	Transferred . .	2/10/1875

Names.	Appointments with Dates.		Remarks with Dates.	
Col. T. Greenaway, 40th N.I.	Commandant .	27/7/1874	Transferred . .	2/10/1875
Lieut.-Col. C. M. White, General Duty	W. O. and Offg. 2nd in command	31/12/1874		
Lieut.-Col. Lavie .	Offg. Wing Officer	1875		
Capt. R. Wilson (Major)	2nd Wing Subaltern	20/8/1875	Transferred . .	1883
	1st Wing Subaltern	24/3/1876		
Major J. Lampen, 10th N.I.	Offg. Wing Officer	2/10/1875	Transferred . .	29/4/1878
	Wing Commander	27/3/1877		
Col. W. A. Riach, 3rd L.I.	Commandant .	2/10/1875	Transferred . .	5/1/1876
Lieut.-Col. W. G. M. Strickland, General Duty	Wing Officer & Offg. 2nd in command	15/12/1875	Transferred . .	27/3/1877
Col. G. Hearn, 12th N.I.	2nd in command .	5/1/1876	Transferred . .	11/9/1876
Lieut.-Col. C. H. Beddek, General Duty	Offg. 2nd in command	22/3/1876	Transferred . .	1876
Surgeon J. G. Collis, 2nd N.I.	Attached as M.O.	3/5/1876	Transferred . .	11/11/1876
Col. W. Rose, 6th N.I.	2nd in command .	10/9/1876	Transferred . .	20/2/1877
Col. F. Dawson, 1st N.I.	Offg. Commandant	11/9/1876	Transferred . .	22/6/1877
	Commandant .	10/11/1876		
Surgeon-Major A. H. Beaman, 36th N.I.	In Medical Charge	11/11/1876	Transferred . .	5/1/1878
Lieut. J. P. James .	Quartermaster .	25/1/1887	Died . . .	...
Lieut.-Col. J. R. S. Henderson, 15th N.I.	2nd in command .	20/2/1877	Transferred . .	4/12/1877
Col. D. G. S. St. J. Grant, 11th N.I.	2nd in command .	4/12/1877	Transferred . .	27/3/1878
Lieut. E. H. Bingham	Offg. Wing Officer	1877	Transferred . .	1877
Col. T. S. Hawks, 37th Gren.	Offg. Commandant	27/6/1877	Transferred . .	12/9/1878
	Commandant .	20/3/1880		
Lieut.-Col. W. G. M. Strickland, 21st N.I.	Offg. Commandant	29/6/1877	Retired . .	18/9/1886
	Offg. Commandant	26/4/1879		
	Commandant .	20/3/1880		
Surg.-Major J. S. Riding, I.M.S.	In Medical Charge	5/1/1878	Transferred . .	11/7/1888
Lieut.-Col. J. L. Benwell, 11th N.I.	Wing Commander	29/4/1878	Transferred . .	6/11/1878
Major W. Anderson, General Duty (Lieut.-Col., Col.)	Offg. Wing Commander	29/4/1878	Retired . .	15/3/1888
	Wing Commander	6/11/1878		
	Commandant .	19/9/1886		
Surg. A. H. Leapingwell, 6th N.I.	In Medical Charge	12/8/1878	Transferred . .	19/11/1880
Col. G. T. Hilliard, 17th N.I.	Commandant .	12/9/1878	Transferred . .	25/4/1879
Col. R. S. Couchman, 19th N.I.	Commandant .	17/6/1879	Transferred . .	16/12/1879
Col. E. F. Waterman, 11th N.I.	Commandant .	26/4/1879	Transferred . .	13/6/1879
Surgeon Barnet, I.M.S.	Offg. Med. Charge	1879		1881
Lieut.-Col. H. H. C. G. Warrington, 29th N.I.	Offg. 2nd in command	26/4/1879	Retired . .	1/1/1886
Col. W. E. White, 9th N.I.	Commandant .	17/12/1879	Transferred . .	20/3/1880

Names.	Appointments with Dates.		Remarks with Dates.	
Surgeon A. H. Williams, 9th M.I.	Medical Charge of Headquarters and Wing	11/10/1879	Transferred . .	12/9/1880
Lieut. H. G. Way, 68th Foot	Offg. Wing Officer	20/11/1880	Transferred . .	18/8/1881
Surgeon-Major W. J. Busteed, I.M.S.	In Medical Charge	17/1/1882	Transferred . .	29/8/1882
Surg. A. G. E. Newland, I.M.S.	In Medical Charge	8/11/1882	Transferred . .	4/3/1883
Lieut. J. E. Preston, 36th N.I. (Captain)	W. O. and Q.M. . Wing Commander	7/6/1882 1/11/1889	Transferred . .	30/4/1890
Lieut. E. C. W. M. Kennedy, York Regt.	Offg. Wing Officer	26/6/1882	Transferred . .	2/11/1882
Lieut. G. A. Welman, 5th N.I. (Captain, Major)	Wing Officer . Adjutant . . Wing Commander 2nd in command .	19/6/1882 10/8/1882 17/4/1891 8/5/1893		
Lieut. J. H. Smith, 2nd Devon Regt. (Captain)	Offg. Wing Officer Quartermaster .	17/7/1882 1/5/1885	Transferred . .	31/3/1896
Lieut. E. D. Ormond, R.M.L.I.	Offg. Wing Officer	17/10/1882	Transferred . .	15/5/1883
Surgeon N. Chatterjie, I.M.S. (Surgeon-Major, Surgeon-Lieut.-Col., Lieut.-Col.)	In Medical Charge	23/11/1882		
Lieut. H. D. Gerrard, 3rd P.L.I.	Wing Officer .	28/12/1882	Transferred . .	22/2/1883
Lieut. C. J. W. Grant, 1st Suffolk Regt.	Offg. Wing Officer	30/8/1883	Transferred . .	21/8/1884
Lieut. H. N. Taylor, 1st Bed. Regt.	Offg. Wing Officer	12/6/1883	Transferred . .	12/7/1884
Lieut. A. J. Richardson, 2nd E. Yorkshire Regt.	Offg. Wing Officer Wing Officer .	10/4/1884 23/10/1885	Transferred . .	24/6/1887
Lieut. H. W. Lowry, 7th N.I.	Wing Officer .	26/6/1884	Transferred . .	4/5/1886
Lieut. C. W. Wilkieson, Cheshire Regt.	Offg. Wing Officer	22/1/1886	Transferred . .	26/2/1886
Lieut. E. W. Carrick, 1st Hants Regt.	Offg. Wing Officer Adjutant . .	13/4/1886 17/1/1890	Transferred . .	24/12/1890
Major W. L. Ranking, G.L.I. (Lieut.-Col., Col.)	2nd in command . Commandant .	19/9/1886 16/4/1888	Transferred . .	8/5/1893
Surg.-Major Youngerman, I.M.S.	Offg. Med. Charge	31/11/1886	Transferred . .	1887
Lieut.-Col. G. Godfrey, 28th M.I.	Offg. 2nd in command	3/12/1886	Transferred . .	22/6/1888
Surg.-Major H. G. Hall, I.M.S.	Offg. Med. Charge	3/12/1886	Transferred . .	1887
Capt. T. F. Leader, 2nd M.I.	Attached . .	3/12/1886	Transferred . .	1/7/1887
Lieut. E. W. L. Holt, 6th M.I.	Attached . .	3/12/1886	Transferred . .	8/11/1887
Lieut. A. H. Allenby, York and Lanc. Regt. (Captain)	Offg. Wing Officer	2/4/1887		
Lieut. R. M. Bell, 19th M.I.	Temp. Attached .	22/7/1887	Transferred . .	14/9/1888
Lieut. J. W. Drever, 19th M.I.	Temp. Attached .	22/7/1887	Transferred . .	4/11/1887

Names.	Appointments with Dates.		Remarks with Dates.	
Lieut. B. Trydell, 9th M.I.	Attached . .	14/9/1888	Transferred . .	30/4/1890
Lieut.-Col. C. H. Sheppard, M.S.C.	2nd in command . Commandant .	23/11/1888 8/5/1893	Transferred . .	8/5/1893
2nd Lieut. R. H. M. Currie, W. Riding Regt.	Offg. Wing Officer	15/3/1890	Transferred . .	4/11/1890
2nd Lieut. W. C. S. Prince, 2nd Middlesex Regt. (Lieut.)	Offg. Wing Officer Quartermaster .	20/4/1890 10/8/1891		
Capt. J. F. Wilson, 8th M.I.	Wing Commander	12/5/1890	Transferred . .	17/4/1891
Lieut. A. Fitz W. Johnson, 12th M.I.	Offg. Wing Officer	12/5/1890	Died at Bangalore	31/10/1891
Lieut. R. P. Jackson, 33rd M.I.	Offg. Wing Officer Quartermaster . Adjutant . .	12/5/1890 10/8/1890 10/8/1891		
Lieut. P. C. Eliott-Lockhart, 1st W. India Regt.	Offg. Wing Officer	14/11/1890	Transferred . .	24/8/1892
Lieut. H. R. Baker, 2nd W. India Regt.	Offg. Wing Officer	1/2/1891		
Lieut. E. Codrington, Wiltshire Regt.	Offg. Wing Officer	22/6/1891	Transferred . .	24/1/1892
2nd Lieut. W. R. B. Colan, 2nd W. York Regt. (Lieut.)	Offg. Wing Officer	14/12/1891	Transferred . .	30/6/1893
Capt. M. E. O'Donoghue, 26th M.I.	Wing Commander	14/7/1892	Transferred . .	9/6/1893
2nd Lieut. J. H. Lawrence-Archer, R.A. (Lieut.)	Offg. Wing Officer	11/10/1892	Transferred . .	22/3/1894
2nd Lieut. G. C. Burn, 2nd Glo'ster Regt. (Lieut.)	Offg. Wing Officer	11/10/1892		
Lieut.-Col. George C. Fenwick, 1st M.P.	Commandant .	12/6/1893	Transferred . .	22/3/1894
Capt. D. S. Lewis, 31st (6th B.B.) M.I.	Wing Commander	22/1/1894	Transferred . .	28/10/1894
Lieut. H. W. Marsden, W. India Regt.	Offg. Wing Officer	30/1/1894	Transferred . .	13/4/1898
Capt. Herbert Lawson, 14th M.I. [Major, Lieut.-Col. (temp.)]	Commandant .	26/5/1894		
2nd Lieut. W. C. Anderson, Devon Regt. (Lieut.)	Offg. Wing Officer	17/9/1894		
Capt. J. A. Loudon, 27th M.I.	Wing Commander	8/11/1894		
Lieut. A. T. Kirkwood, 29th (7th B.B.) M.I.	Offg. Wing Officer	12/7/1895	Transferred . .	8/11/1897
Lieut. W. C. T. G. G. Plant, 20th Hussars	Offg. Wing Officer	18/2/1898		
Surg.-Major Kanga, 9th M.I.	Offg. Med. Charge	1/4/1898	Transferred . .	1/10/1898

(4) NAMES OF OFFICERS KILLED AND WOUNDED.

Name.	Place.		Date.
Captain Hugh Robert Alcock	Pondicherry	Wounded	Oct. 9, 1778
Ensign Henry Weber .	Pollilloor	,,	Aug. 27, 1781
,,	Carroor	,,	Mar. 21, 1783
Ensign Robert Gahagan .	Annagudi	,,	Feb. 18, 1782
Ensign Haywood . .	,,	,,	,, ,,
Major Colin Campbell .	Sultanpet	Killed	April 6, 1799
Captain Alexander Stewart	Wynaad	Wounded	1811
Lieutenant Inverarity .	Cotiaddy Pass	Severely wounded	April 4, 1812
Captain George Hunter .	,,	Severely wounded in the right arm by an arrow. His arm being described as nearly useless, he was granted a wound pension.	,, ,,
,,	Mahedpore	Severely wounded	Dec. 21, 1817
Lieutenant William J. D. Glen	,,	Killed	,, ,,
Lieutenant John Jones .	,,	Wounded	,, ,,
Lieutenant Joseph Standiver Sherman	Fodgaghee Stockade	Slightly wounded	Mar. 25, 1825
Captain J. E. Preston .	Lamaing	Severely wounded	1886

(5) NAMES OF OFFICERS AND OTHERS WHO HAVE DISTINGUISHED THEMSELVES.

CAPTAIN J. OLIVER took part in the capture of Pondicherry on the 9th October 1778, and shared in the prize money (*Madras Military Letter*, dated 21st October 1806). Led the storming party at the capture of Rayacottah, 20th July 1791. Captured, with detachment 13th Battalion, Valoidum Naik, Polygar of Pylney, in his fort at Baulsamoodrum in the Madura district, 7th October 1795. He and his detachment were thanked in General Orders.

JEMADAR BOODH SINGH, the jemadar of Captain Oliver's party, was promoted subadar and presented with a gold medal, having on one side the words "Courage and Fidelity," and on the other, "By Government, 7th October 1795."

LIEUTENANT ARTHUR MOLESWORTH "served during the whole campaign against Tippoo Sultan. Was present at the battle of Mallavelly under General Harris. Served during the whole siege, and was in command of the light companies of the 2nd Battalion, 3rd Regiment, in the storming party of Seringapatam, 4th May 1799, and had the honour of receiving the medal presented to commemorate that day" (*East India Military Calendar* for 1823–24).

SUBADAR SELAZEE ROW, in consideration of his long, active, and zealous services, was pensioned on his full pay, and was awarded a monthly allowance of Ten Pagodas for a Palanquin, with effect from the 1st July 1800 (*G.O.G.*, dated 3rd July 1800).

LIEUTENANT-COLONEL BARRY CLOSE was highly commended by the Honourable the Court of Directors, while serving as Adjutant-Genera of the Army, and as a testimony of their esteem for the distinguished services rendered by him in the war against Tippoo Sultan, was presented with a sword of honour (*G.O.G.*, dated 25th December 1800).

LIEUTENANT GEORGE HUNTER lost the whole of his baggage on the 28th May 1801, whilst attached to the Poona Subsidiary Force and detached to a distant station at a period of great emergency. The misfortune which befel him in the plunder of his baggage by armed band of banditti was "in a considerable degree ascribed to the zeal which he evinced in the execution of the orders which he had received" (*Finance Department Letter*, dated Madras, 8th September 1805). He received Rs.1732 as compensation.

MAJOR JAMES WELSH distinguished himself at the capture of the Arambooly Lines on the 10th February 1809. His services on that occasion are thus recorded :—" The storming party, consisting of . . . and nine companies selected from both battalions of the 3rd Regiment, under twenty-five officers, assembled at his post in advance at 8 p.m. ; he explained his plans to all the officers, and then set forward on expedition, which appeared so rash even to those whom he was leading, that ere they had proceeded far he formed a forlorn hope of volunteer Europeans, and headed it himself. The night was extremely dark, and though the total distance was within three miles, it took eight and a half hours to reach the works at the summit of the hill, scrambling through very thick jungles, into deep ravines, and over rugged rocks. At 4.30 a.m., 10th February, he found himself with another officer with the forlorn hope alone at the foot of a stone wall twelve feet high, having been directed to it by the enemy's patrols, who had just passed by with numerous lights. This was the moment for decision ; he seized it, and with twenty resolute followers entered the place, climbing upon one another's shoulders, etc. Nearly at the same time the head of the storming party reached a part of the wall about 200 yards lower down, ladders were applied, and the whole works carried before daylight, under a very heavy fire of cannons and musketry, but which did little execution. He received public thanks in Detachment and General Orders. The works were named after him (though subsequently destroyed). The surviving officers of the storming party presented him with an address and piece of plate, on which is inscribed their admiration of his conduct" (*East India Military Calendar* for 1823–1824).

CAPTAIN LUCAS commanded five companies of the regiment at the capture of the Arambooly Lines on 10th February 1809, and was mentioned in despatches.

COLONEL HENRY WEBBER, whilst commanding at Ternate in 1801, repeatedly received thanks for his conduct in reconciling the discordant interests of the Malay princes of Ternate, Tidore, and Bachian ; and on his being ordered to deliver over the island to the Dutch, he received the public thanks of the Dutch Governor who relieved him, in the name of the Sultan and inhabitants of Ternate. (2) Also received the thanks of Government whilst employed in 1812 in reducing to obedience the rebellious Polygars in the Wynaad. (3) Received the approbation of Government for the readiness with which he complied with the application of the Resident at Hyderabad for the aid of a detachment (*Political Letter*, dated 16th August 1820).

SUBADAR SHEICK ESMAEL was presented with a palanquin and 20 pagodas per month for the support of that equipage, "in consideration of his long and faithful services." On his demise his nearest heir was ordered to be pensioned on the half-pay of a subadar of infantry (*G.O.G.*, dated 12th November 1811).

LIEUTENANT GEORGE DODDS was present at the capture of Mahedpore in 1817, and shared the prize money obtained during the war against the Pindarees.

LIEUTENANT JAMES BRIGGS shared the prize money obtained during the war against the Pindarees and certain of the Mahratta States in 1817–18. (2) On the 8th October 1824 was appointed to the Survey Branch of the Q.M.G.'s Department, and was present with the unsuccessful detachment of Lieutenant-Colonel Smith, when he "zealously volunteered his services to conduct a party in returning to secure some ladders to assail the pagoda at Keykloo, when he was attacked by thirty or forty Burmans, who rushed upon him with drawn knives, and from whom he only escaped by jumping down a deep ravine." He received the thanks of the Commanding Officer for his services in the Q.M.G.'s Department on that occasion (*London Gazette*, dated 25th November 1825).

LIEUTENANT JOSEPH STANDIVER SHERMAN'S services in escalading the stockade near Fodgaghee, when he was slightly wounded in the engagement, were brought to notice by Lieutenant-Colonel Smith (*London Gazette*, dated 25th March 1825).

LIEUTENANT JOHN SHIEL'S services "in leading on his men on Major Wahab's being wounded in the unsuccessful attack by Lieutenant-Colonel Smith's detachment upon the pagoda at Keykloo" were highly commended in General Orders. He received a donation of 6 months' batta.

LIEUTENANT GEORGE WILLIAM WATSON served in the Burmese War of 1825, and received a donation of 6 months' batta.

LIEUTENANT JAMES FRASER LESLIE. When placed in civil charge of Mergui after Captain Buxton's retreat, Major Burney, the British Resident at Ava, reported that "his knowledge of the Burmese language, intimate acquaintance with the place and people from former residence there, and particularly whose mild and prepossessing manner renders him well qualified to restore everything to its former fortunes at Mergui, with the inhabitants of which place he is a great favourite" (*Madras Military Department Letter*, dated 11th December 1829).

LIEUTENANT JOSIAH SMITH carried a despatch on the 26th March 1837 to Colonel Burney, the British Resident at Ava, and was reported by the Brigadier-General Commanding the District "to have performed an important and difficult service with great zeal, promptitude, and intelligence." The Governor-General of India "considered the service performed to be highly creditable to him, of having effected, at an important juncture, a communication with the Resident at Ava" (*India Office Records*).

SUBADAR GOVINDOO received the 1st class of the Order of British India, with the title of Sirdar Bahadur, for conspicuous gallantry in the defence of his post at Ky-ka-ti in Martaban against a large body of rebels, headed by the chief "Moung Thelah," on the 4th March 1858.

NO. 1509, PRIVATE MAHOMED HOOSMAN, particularly distinguished himself on the 21st May 1858, in the pursuit and capture of certain escaped convicts at Mergui in Burma.

CAPTAIN J. E. PRESTON, D.S.O., in the Burmese Expedition of 1885 to 1888, at the night affair at "Lamaing" and other actions, was twice mentioned in despatches (*London Gazette* of 22nd June 1886, and 2nd September 1887). Received the D.S.O.

FIRST GRADE HOSPITAL ASSISTANT JENARDEN SINGH specially promoted to Senior Hospital Assistant for services in Burma, and afterwards created a "Rai Bahadur."

(6) LIST OF NATIVE OFFICERS WHO HAVE SERVED IN THE REGIMENT.

(Names spelt as they appeared in General Orders.)

No.	Rank.	Names.	Jemadar.	Native-Adj.	Subadar.	Subr.-Maj. (*The rank of Subadar-Major was first introduced into the native army on the 2nd Feb. 1819.*)
	Havildar	Selazee Row	1790	...	1795	...
	,,	Bood Singh	1795	...	...	...
	,,	Shaik Jaffer	1800	...	...	...
	,,	Ramlingam Naick	1800	...	...	...
	,,	Shaik Ismael	1800	...	...	...
	,,	Sheshgerry	1802	...	1/1/1808	...
	,,	Ismael Sahib	1802	...	1/1/1808	...
	,,	Bavamy Sing	1802	...	1/1/1808	...
	,,	Mucktoom Saheb	1802	...	1/1/1808	...
	,,	Ebram Beg	1802	...	1/1/1810	...
	,,	Shaik Esoph	1803	...	1/1/1810	...
	,,	Mootoo Sawmy	1803	...	1/1/1810	...
	,,	Shaik Ahmed	10/10/1804	...	...	...
	,,	Shaik Adam	10/10/1804	...	...	...
	,,	Shaik Ibrahim	10/10/1804	2/3/1808	...	...
	,,	Mahomed Ally	1/3/1805	8/1/1811	...	...
	,,	Harry Sing	1/1/1808	...	...	...
	,,	Goolam O'Deen	1/1/1808	...	...	...
	,,	Mahomed Ally	12/9/1808	...	...	...
	Hav.-Maj.	Surgh Cawn	1/1/1810	...	...	...
	Drill-Hdr.	Mootoo Sawmy	1/1/1810	...	...	...
	Havildar	Noor Mahomed	1/12/1810	...	...	...
	Subadar	Shaik Adam (from 1st Bn.)	...	...	31/12/1813	...
	Havildar	Mahomed Esoph	1/2/1819	...	...	...
	,,	Vencataram	...	...	...	...
	,,	Shaik Secunder	?	...	...	...
	,,	Shumsheeroodeen	?	...	?	...
	,,	Jaffer Homed	?	...	?	...
	,,	Shaik Curreem	?	...	?	...
	,,	Soobanjee Row	?	...	?	...
	,,	Mahomed Ally	?	...	?	...
	,,	Surjea Cawn	?	...	?	...
	,,	Ramjee	?	...	...	...
	,,	Shaik Esmaul	?	...	?	?
	,,	Ram Singh	?	...	...	...
	,,	Gungaram	...	...	...	...
	,,	Ebram Beg	...	...	...	1/1/1827
	,,	Mahomed Esoph	?	...	1/1/1824	...
	,,	Mootoo	?	..	?	...
	,,	Shaik Hyder	?	..	1/1/1824	...
	,,	Shaik Luttif	?	1/1/1824	1/1/1827	...
	,,	Shaik Emom	1/1/1824	27/4/1827	23/4/1828	...
	,,	Venketran	...	...	1/1/1826	...
	,,	Shaik Madar	...	...	21/1/1826	...
	,,	Peer Mahomed	1/1/1824	...	20/8/1828	...
	,,	Shaik Modeen	...	...	1/1/1827	...
	,,	Hurry Singh	1/1/1824	...	2/2/1832	...
	,,	Syed Hussein	...	...	...	8/3/1831
	,,	Gungaram	1/1/1824	...	18/5/1833	...
	,,	Bawboo Cawn	1/3/1825	...	24/12/1834	...
	,,	Shaik Amud	1/1/1826	...	...	...

No.	Rank.	Names.	Jemadar.	Native-Adj.	Subadar.	Subr.-Maj.
	Havildar	Ramasamy	1/1/1826	...	...	...
	,,	Shaik Hoosain	?	...	?	...
	,,	Shaik Homed	?	...	9/8/1836	28/4/1840
	,,	Ramasawmy	21/1/1826	...	...	...
	,,	Shaik Adam	1/1/1827	...	...	...
	,,	Bram Khawn	1/1/1827	...	...	...
	,,	Mahomed Yath	23/4/1828	...	13/8/1837	...
	,,	Shaik Meean	15/8/1828	20/9/1828	23/2/1838	...
	,,	Said Hoossman	20/8/1828	...	?	...
	,,	Shaik Moodeen	8/3/1829	...	22/1/1839	...
	Hav.-Maj.	Shaik Mustan	2/2/1832	23/2/1838	5/2/1839	20/2/1844
	Havildar	Narrain Singh	18/5/1833	5/4/1839	16/8/1839	...
	Hav.-Maj.	Jey Singh	18/5/1833	...	...	...
	Havildar	Laul Mohun	24/12/1834	6/12/1839	14/2/1840	...
	,,	Venketasamy	?	5/6/1840	...	...
	,,	Narrainah	9/8/1836	...	28/4/1840	...
	,,	Baboo Khan	13/8/1837	...	1/3/1842	...
	,,	Gunga Sing	23/2/1838	...	...	...
	,,	Shaik Lattief	22/1/1839	...	20/2/1844	...
	,,	Shaik Jee	22/1/1839	...	12/3/1844	29/4/1856
	Hav.-Maj.	Timmapah	5/2/1839	...	24/9/1844	...
	Havildar	Siddy Hoosain	16/8/1839	...	28/3/1845	...
	,,	Venketasawmy	16/8/1839	...	29/8/1845	...
	,,	Rhyman Beg	14/2/1840	...	29/8/1845	...
	Hav.-Maj.	Govindoo	28/4/1840	29/8/1845	19/2/1847	...
27	,,	Venketachellum	1/3/1842	...	...	...
165	Havildar	Sied Fiddooally	1/3/1842	...	2/3/1847	...
92	,,	Shaik Ally	20/2/1844	...	...	...
206	Hav.-Maj.	Shaik Homed	12/3/1844	...	14/2/1847	...
	Havildar	Chatterpermaul	...	...	24/4/1855	...
189	,,	Narsoo	24/9/1844	...	17/7/1855	...
266	Hav.-Maj.	Mootoosawmy	28/3/1845	...	29/4/1856	...
381	Havildar	Secunder Cawn	29/8/1845	23/3/1847	...	...
209	,,	Mootialoo	29/8/1845	...	1/8/1857	...
81	,,	Narrainah	23/12/1845	...	...	...
203	,,	Cassylingum	19/2/1847	...	22/9/1857	...
201	Hav.-Maj.	Mootooveeroo	2/3/1847	...	...	...
94	Havildar	Shaik Moideen	14/12/1847	...	7/10/1857	...
405	Hav.-Maj.	Rungiah	24/7/1848	...	2/3/1858	...
123	Havildar	Shaik Meeran	4/2/1852	...	...	...
465	,,	Sied Lutteef	23/12/1852	...	18/8/1858	...
162	,,	Chinnatombyroyah	24/4/1855	...	...	...
503	,,	Soobiah	24/4/1855	...	14/10/1858	...
	Subadar-Major	Veerasawmy (from late 42nd N.I.)	8/10/1844	...	26/4/1853	5/2/1864
447	Havildar	Shaik Homed	17/7/1855	...	3/5/1859	30/6/1864
339	,,	Shaik Bram	29/4/1856	...	9/3/1860	...
	Subadar	Meerzeuratoola Beg (from 45th N.I.)	16/5/1854	...	29/4/1860	...
517	Drill Hav.	Shaik Emaum	1/8/1857	...	14/3/1862	...
	Havildar	Syed Cosseem	...	...	14/3/1862	...
496	,,	Boorandeen	22/9/1857	...	13/6/1864	...
518	Hav.-Maj.	Yacoob Khan	7/10/1857	18/8/1859	30/6/1864	16/9/1874
	Jemadar	Mahomed Cassim (45th N.I.)	7/9/1858	...	30/6/1864	...
384	Cr. Hav.	Sied Ally	2/3/1858	...	...	...
375	Cr. Hav.	Mahomed Shereeff	18/8/1858	...	...	...
865	Havildar	Sied Moonadeen	14/10/1858	...	30/6/1864	11/3/1877
	Jemadar	Abdool Khadir (from 45th N.I.)	29/3/1859	...	15/2/1869	...
415	Havildar	Sied Jemallodeen	3/5/1859	...	...	...
491	,,	Sied Goodoo	9/3/1860	...	...	...
593	,,	Rungasawmy	14/3/1862	...	...	...
718	,,	Soobiah	14/3/1862	...	15/11/1868	16/8/1878

No.	Rank.	Names.	Jemadar.	Native-Adj.	Subadar.	Subr.-Maj.
552	Havildar	Mooneah	13/6/1864	...	15/2/1869	...
1245	,,	Shaik Daood	30/6/1864	23/7/1864	15/2/1869	1/12/1879
609	,,	Abdool Cawder	30/6/1864	...	1/5/1873	...
745	,,	Shaik Homed	17/3/1866	...	1/5/1873	...
88	,,	Narrainsawmy	15/11/1868	...	16/9/1874	...
36	,,	Shaik Jemall	15/2/1869	...	16/3/1876	...
86	,,	Sied Huneef	15/2/1869	...	...	...
111	,,	Venketsawmy	15/2/1869	...	11/2/1878	...
181	,,	Lutchmiah	15/2/1869	27/3/1869	11/3/1877	16/5/1881
168	,,	Narrainsawmy	15/2/1869	...	...	...
73	,,	Hoossain Khan	1/5/1873	...	...	...
82	,,	Permalloo	1/5/1873	...	...	...
58	,,	Meer Sudrodeen	16/9/1874	...	...	...
45	,,	Sheik Burray	11/11/1874	...	...	...
114	,,	Mahomed Cassim	16/3/1876	...	16/8/1878	...
115	,,	Mucktoom Khan	11/3/1877	15/3/1877	16/8/1878	...
249	,,	Venketasawmy	11/2/1878	...	1/12/1879	...
245	,,	Ramasawmy	16/8/1878	16/8/1878	1/12/1879	1/12/1882
300	,,	Ramasawmy	16/8/1878	...	16/5/1881	...
	,,	Gholam Moideen	1/12/1878	...	16/5/1881	16/4/1886
288	,,	Jemall Begh	1/1/1879	...	...	...
189	,,	Chinnasawmy	1/1/1879	...	6/11/1881	...
169	,,	Moideen Khan	1/1/1879	...	...	...
384	,,	Ramanjaloo	22/11/1879	...	...	...
500	,,	Syed Bahavadeen	1/12/1879	26/1/1880	1/11/1882	7/5/1887
256	,,	Sheikh Homed	1/12/1879	...	1/11/1882	...
	,,	Sevadial	21/11/1880	...	1/12/1882	...
	Hav.-Maj.	Seeshachellum	16/5/1881	18/12/1882	21/12/1884	26/4/1890
	Havildar	Mahomed Cader	16/5/1881	...	16/4/1886	...
	,,	Audiah	6/11/1881	...	...	...
	,,	Jeanah	1/5/1882	...	16/4/1886	...
	,,	Peer Homed	1/11/1882	...	1/9/1886	...
	Hav.-Maj.	Shaik Furreed	1/11/1882	2/3/1885	7/5/1887	...
	Havildar	Venkutramanjiah	1/12/1882	...	...	...
	,,	Mootialsawmy	6/3/1883	...	...	...
	,,	Cunniah	21/12/1884	..	30/6/1887	...
665	Hav.-Maj.	Jaganad Sing	16/4/1886	...	11/4/1888	6/7/1895
	Havildar	Syed Mahmood	16/4/1886	...	...	...
	,,	Narrainsawmy	16/4/1886	...	5/12/1888	...
	,,	Mahomed Cassim	1/9/1886	...	6/4/1890	...
	,,	Mahomed Ackber	2/10/1886	...	...	...
	,,	Kullian Singh	7/5/1887	...	...	...
747	Hav.-Maj.	Khader Moideen	30/6/1887	30/9/1887	26/4/1890	...
	Havildar	Mootoonaidoo	1/1/1888	...	2/6/1893	...
1000	,,	Saiyid Umar	20/3/1888	1/7/1890	26/12/1893	...
	,,	Dost Muhammad Khan	11/4/1888	...	...	...
	Hav.-Maj.	Pitchamootoo	5/12/1888	...	...	...
1002	Havildar	Chedanum	6/4/1890	...	24/1/1895	...
1001	,,	Veragooloo	26/4/1890	...	6/7/1895	...
1022	,,	Sikandar Khan	1/3/1891	...	1/11/1895	...
1070	,,	Abdur Razzak	1/1/1892	26/12/1893	1/1/1897	...
1031	,,	Muhammad Ghaus	9/8/1892	...	...	...
1075	,,	Abdur Rahman Khan	2/6/1893	...	...	...
1192	,,	Audinarayadu	24/12/1893	...	1/3/1897	...
1228	,,	Sayyid Muhiyuddin	24/1/1895	...	...	...
1199	,,	Runganayakulu	25/5/1895	...	...	...
1415	,,	Mannikkam Mudali	6/7/1895	1/1/1897	24/1/1898	...
1434	,,	Azhagarsami	1/11/1895	...	...	...
1238	,,	Amir Beg	1/1/1897	...	...	...
1325	,,	Venkatasami	1/3/1897	24/1/1898	..	...
2250	,,	Ponnusami	24/1/1898	...	...	...
	,,	Abdur Rahmon Khan	1898	...	...	...

(7) ORDER OF BRITISH INDIA.

1837 The "Order of British India" was established on the 1st June 1837 by the Court of Directors, as one of the means for "improving the condition of the native soldiery," and is given to subadars and jemadars for long and honourable service. The members of the Order are divided in two classes: the first class receiving an allowance of two rupees a day each, in addition to allowances or retiring pensions, with the title of "Sirdar Bahadur"; and the second class an allowance of one rupee a day each, in addition to their usual allowances and pensions, with the title of "Bahadur."

List of Officers who have been Granted the Order of British India.

Names.	2nd Class.	1st Class.
Subadar-Major Sayyid Husain	Bahadur, 26/1/1839	
Subadar Govindoo	...	Sirdar Bahadur, 15/9/1857
Subadar-Major Yacub Khan	...	Sirdar Bahadur, 1/1/1877
Subadar Narrainsawmy	Bahadur, 5/7/1877	Sirdar Bahadur, 17/4/1881
Subadar-Major Sayyid Bahaud-din	Bahadur, 8/6/1888	
Subadar-Major Jagannath Singh	[1]Bahadur, 13/8/1897	

[1] In commemoration of the sixtieth year of Her Majesty's reign.

(8) GOOD CONDUCT MEDALS.

On the 31st January 1888, the Governor-General of India in Council announced that the sanction of Her Majesty's Government had been accorded to the grant of good conduct medals with annuities and gratuities as follows :—To each Native Infantry Regiment a silver medal, inscribed "For Meritorious Service," with annuity of Rs.25 to a havildar of decidedly meritorious conduct, the annuity to continue after the havildar is pensioned, if recommended by the Commandant of his regiment.

1888

To each Native Infantry Regiment yearly two silver medals, inscribed "For Long Service and Good Conduct," to two men of the rank and file who are the most deserving of those men who have qualified for the honour and are eligible for it.

The grant of these good conduct medals as a reward for a long course of irreproachable conduct is announced in Orders, in order that every man who obtains one may be held up as an object of respect and emulation to the non-commissioned officers and soldiers of the corps in which he has served. The medals are transmitted to the Commandant for presentation by the Commanding Officer of the station, to be worn by the recipient as an honourable testimony of the approbation in which his conduct is held.

The under-mentioned non-commissioned officers of the regiment have been granted medals, inscribed "For Meritorious Service," with annuity, under the provisions of Clause 115, I.A.C. of 1888 :—

	Date.	Presented by
No. 531, Havildar Arokeum	1/3/1889	Col. W. L. Ranking, Commandant
No. 784, Havildar-Major Ramasawmy	27/6/1894	Captain H. Lawson, Commandant
[1]No. 1144, Band Havildar Robert Paul	10/12/1897	Lt.-Col. H. Lawson, Commandant

The under-mentioned men of the regiment have been granted medals, inscribed "For Long Service and Good Conduct," with gratuity, under the provisions of Clause 115, I. A.C. of 1888:—

No.	Name.	For the Year ending	Presented by
369	Private Sayyid Buddhan . .	31/3/1889	Colonel W. L. Ranking, Commandant.
410	Private Affanah	,,	,,
489	Private Sunnasse . . .	31/3/1890	,,
674	Private Shaik Hamid . .	,,	,,
388	Private Shaik Husain . .	31/3/1891	,,
910	Private Soobarayadoo . .	,,	,,
411	Private Appiah . . .	31/3/1892	Brigadier-General H. M. Bengough, C.B.
522	Private Sayyid Haidar . .	,,	,,
955	Private Venketrajoo . . .	1/3/1893	Colonel W. L. Ranking, Commandant.
970	Private Ramanah . . .	,,	,,
632	Private Muhammad Yakub .	31/3/1894	Lieutenant-Colonel G. C. Fenwick, Commandant.
788	Private Sayyid Kasim . .	31/3/1895	Captain H. Lawson, Commandant.
813	Private Sayyid Kasim . .	,,	,,
775	Rungasawmi	31/3/1896	,,
1379	Wazir Khan	,,	,,
1040	Private Narayanasami . .	31/3/1898	Lt.-Col. H. Lawson, Com.
1047	Private Vembuli . . .	,,	,,
996	Private George Bradford [2] .	10 12 1897	,,
2241	Private Ramasami [2] . . .	,,	,,
	Band Naick Solomon . .	31/3/1898	,,

Drummer John Bearder was awarded a gratuity medal for long service on the 26th August 1863.

A silver medal for "Long Service and Good Conduct," together with a gratuity of £5, was awarded to Drum-Major John Charles M'Connell of the regiment, on the 13th July 1882, for the year 1881–82.

[1] In commemoration of the sixtieth year of Her Majesty's reign. The medal with gratuity only was awarded.

[2] In commemoration of the sixtieth year of Her Majesty's reign.

(9) RETURN SHOWING EFFICIENCY OF SIGNALLERS OF THE REGIMENT.

Year.	Order of Merit.		No. of Regimental Signallers examined.	Small Flag. Pairs.		Heliograph. Pairs.		Heliograph. Assistant Instructors.		Lamp. Assistant Instructors.		Small Flag, Service Message.	Accuracy.						No. of Efficient Signallers.	Bonus to which entitled.	
	N.I. Regiments in India.	N.I. Regiments in Madras.		Average rate of sending.	Percentage of Letters read correctly.	Average rate of sending.	Percentage of Letters read correctly.	Average rate of sending.	Percentage of Letters read correctly.	Average rate of sending.	Percentage of Letters read correctly.		Figure of Merit.	Flag, Pairs.	Heliograph, Pairs.	Heliograph Assistant Instructors.	Lamps.	In all four.		No. of Signallers.	Amount each.
1894–95	4	3	24	12.76	97.97	13.04	92.20	14.81	96.37	13.63	98.18	18.00	456.96	...	...	A.	...	...	23	8	Rs. 5
1895–96	19	9	23	12.76	96.25	14.63	88.57	15.78	98.18	15.58	96.68	18.33	456.76	A.	V. I.	V. A.	A.	F. A.	23	8	Rs. 5
1896–97	20	10	25	13.63	96.15	13.95	94.97	15.38	96.25	15.00	97.54	20.00	462.87	A.	F. A.	A.	V. A.	A.	21	8	Rs. 5

(10) CERTIFICATES OF EDUCATION AWARDED AT THE REGIMENTAL SCHOOL.

Year.	1st Class.	2nd Class.	3rd Class.	4th Class.	
1880	Nil	Nil	Nil	Nil	
1881	Nil	Nil	5	32	
1882	Nil	Nil	11	13	
1883	Nil	5	6	25	
1884	Nil	Nil	4	17	
1885	Nil	Nil	4	13	
1886	Nil	Nil	10	19	
1887	Nil	Nil	Nil	Nil	On foreign service.
1888	Nil	Nil	Nil	Nil	
1889	Nil	1	4	9	
1890	Nil	Nil	19	8	
1891	Nil	2	24	21	
1892	Nil	8	15	10	
1893	Nil	Nil	13	17	
1894	Nil	6	23	29	
1895	Nil	Nil	Nil	Nil	
1896	Nil	3	19	30	
1897	Nil	Nil	3	17	

(11) MEDICAL STATISTICS OF THE REGIMENT.

Year.	Station.	Average Strength.	Actual number admitted in Hospital.	Average daily sick.	Death in and out of Hospital.	Sick per Mille.	Death per Mille.	Invalided during the Year.	Sent on sick leave.
1887	Pyinmana and Yemethin .	803	691	27	37	33	46	...	77
1888	Pyawbwé, headquarters and right wing	483	451	26	31	54	64	2	83
1889	Meiktila, headquarters and right wing	538	443	20	22	37	41	1	47
1890	Bangalore (arrived 26/3/90) .	778	551	16	16	21	21	...	29
1891	Bangalore	817	464	14	7	17	9	23	19
1892	Bangalore, arrived at Cannanore, 15/12/92	838	431	14	5	17	6	...	16
1893	Cannanore	825	363	16	6	19	7	5	3
1894	Cannanore	814	553	23	...	28	...	22	10
1895	Cannanore, arrived at Thayetmyo, 9/11/95	821	436	20	2	24	2	28	16
1896	Thayetmyo	809	627	23	18	28	22	12	22

MEDICAL STATISTICS OF THE REGIMENT IN BURMA FROM 9TH NOVEMBER 1895 TO JANUARY 1898.

Headquarters, Thayetmyo, from 9th November 1895 to 31st December 1897.
Average strength of all native ranks, 540.
Detachment, Mindat Sakan, Chin Hills, from 24th October 1895 to January 1898.
Average strength of all native ranks, 160.

No. of men invalided from Mindat Sakan to Thayetmyo		69
No. of men sent to the M.I. Depôt Pallaveram for disposal on medical grounds		104
No. of men granted sick leave from Thayetmyo to India		24
Nature of disease—Ague	28	
,, Dysentry	26	
,, Rheumatism	15	
,, Malarial cacheuca . .	18	
,, Pneumonia	13	
,, Secondary syphilis . .	14	
,, Debility	14	
,, Diarrhœa	8	
,, Melancholia	3	
,, Other causes	58	197
No. of deaths at Thayetmyo		8
No. of deaths at Mindat Sakan		12
Nature of disease—Dysentry	6	
,, Beri-Beri	4	
,, Other causes	10	20

PART VIII

STATIONS OF THE REGIMENT

VIII

STATIONS OF THE REGIMENT

Name of Station.	From	To	Remarks.
Madras	/12/1776	1778	...
Pondicherry	8/8/1778	1780	...
Tanjore	/7/1780	1781	...
Regiment employed on service at			*The two grenadier companies were detached from the regiment at Tanjore on the 24th July 1780, and served with the Trichinopoly Detachment at*
Tricatapully	/8/1781		Chittapet Fort — 6/9/1780
Pootoocottah	3/8/1781		Great & Little Mounts — /9/1780
Manargudi	8/9/1781		Chingleput, Wandiwash, Permacoil — 17/1/1781
Mahadavypatam	16/9/1781		Pondicherry — /1/1781
Alangudi	30/9/1781		Tiruwadi — /4/1781
Nagore	17/10/1781		Chillumbrum — /4/1781
Negapatam	29/10/1781	Detacht.	Porto Novo — 1/7/1781
Trincomalee	5/1/1782	Volunteers.	Tripassoor Fort — 20/8/1781
Fort Osnaburg	11/1/1782		Pollilloor — 27/8/1781
Annagudi	18/2/1782	300 men	Poloor Fort — /9/1781
Trichinopoly	/3/1783		Sholinghur — 27/9/1781
Caroor	21/3/1783		Paliput — 26/10/1781
Avaracoorchy	10/4/1783		Vellore — 3/11/1781
Dindigul	4/5/1783		Chittoor — 10/11/1781
Darapooram	2/6/1783		Tripassoor — 22/11/1781
Mellore	2/8/1783		Madras — /12/1781
Shevagunga	4/8/1783		Arnee — /4/1782
Panjalumcoorchy	12/8/1783		Madras — 18/6/1782
Shevagherry	/9/1783		Neddingul — 13/2/1783
Dindigul	23/9/1783		Wandiwash, Corangooly — /4/1783
Pulney	16/10/1783		Vellore — 21/4/1783
Cumalum Fort, Chucklegherry, Annamully	/10/1783		Cuddalore — 7/6/1783
Palghautcherry	5/11/1783		Rejoined Regiment — 23/9/1783
Coimbatore	26/11/1783		
Caroor, Dindigul, or Coranoor	1784	1785	
Trichinopoly	1786	/4/1789	
Regiment employed on service at			
Island of Vypeen	/4/1789		
Palghautcherry	1790		
Coimbatore	/9/1790		
Poolanhully	17/11/1790		
Vellore	11/1/1791		
Colar	28/2/1791		
Ooscottah	2/3/1791		
Bangalore	5/3/1791	4/5/1791	
Near Seringapatam	/5/1791	6/6/1791	
Hooliadroog Fort	/6/1791		
Bangalore	11/7/1791	13/7/1791	

Name of Station.	From	To	Remarks.
Regiment employed on service at: Oossoor	15/7/1791		
Anchittydroog			
Neilgherry	/7/1791		
Ruttungherry			
Royacottah	20/7/1791		
Kinchillydroog	/7/1791		
Oodiadroog	/7/1791		
Rahmanghur	14/9/1791		
Nundidroog	22/9/1791		
Cummuldroog	/9/1791		
Savandroog	15/12/1791	/3/1792	
Dindigul	/3/1792	1796	
Regiment on service at: Baulsamoodrum	7/10/1795		Detachment
Vellore	/2/1779	14/2/1799	
Oodiadroog	/3/1799		
Mallavelly	27/3/1799		
Near Seringapatam	5/4/1799	4/5/1799	
Seringapatam	5/5/1799	7/2/1803	
Regiment on service: Eygoor	30/3/1800		5 companies
Arrakaira	2/4/1800		,,
Bullum		/6/1800	
Hurryhur		9/3/1803	
Poona	22/3/1803	/8/1803	
Ahmednugger	9/8/1803	/3/1806	
On service: Korjet Corygaum	/10/1803		4 companies
Umber	31/10/1803		3 companies
Trichinopoly	/4/1806	/12/1808	
Regiment on service: Nagercoil			
Cotaur	17/2/1809		Flank companies
Oodagherry Fort			
Papanaveram	19/2/1809		,,
Colachee	23/2/1809		
Trevandrum	28/2/1809		
Bednore	/8/1809		Detachment at Bangalore
Nundydroog	1809	1809	
Seringapatam	1810	1810	
Cannanore	1810	25/4/1814	
On service: Manantoddy	1810	1814	2 companies
Sultan's Battery	1810	1814	30 men
Vellore	3/6/1814	4/6/1814	
Madras		3/3/1816	
Amaravatty	24/4/1816	18/8/1816	
Ongole	28/8/1816	16/11/1816	
Datchapillay	22/11/1816	10/3/1817	
Humpsaghur	?	?	
Marsulipatam	?	18/10/1817	
Guntoor and the Palnad	?	/3/1818	
Bellary	25/7/1818	9/3/1819	
On service: Baddamy	1819	1819	
Huttony	1819	1819	
Bijapur	25/5/1819	30/11/1819	
Kaladgi	11/2/1820	18/10/1821	
Sattarah	4/11/1821	18/4/1822	
Kaladgi	/5/1822	/3/1823	
Bellary	1/3/1823	30/3/1824	
Gooty	3/4/1824	16/12/1826	Detachment at Cuddapah
Secunderabad	7/1/1827	9/6/1829	
Pallaveram	1829	1829	
Cuddapah	17/7/1829	7/5/1832	
Vellore	25/5/1832	20/6/1835	

Name of Station.	From	To	Remarks.
Madras	29/6/1835	7/10/1835	
Maulmain	2/11/1835	28/4/1839	
Pallaveram	25/5/1839	17/8/1839	
Vellore	23/8/1839	19/12/1840	
Samulcottah	9/2/1841	12/10/1843	Detachments at Ellore and Masulipatam
Chicacole	1/11/1843	20/3/1845	
Secunderabad	8/5/1845	15/1/1849	
Cuddapah	10/2/1849	1/9/1851	
Trichinopoly	6/10/1851	3/10/1854 28/1/1855	Headqrs. and Right Wing Remainder
Palamcottah	23/10/1854 17/2/1855	17/12/1856	
Maulmain	16/2/1857 14/3/1857	24/1/1860	Detachments at Sittang, Mergui, Penoung, Thanoi, and Amherst
Trichinopoly	17/1/1860	23/12/1864 27/5/1864	Left Wing
Cannanore	10/6/1864 13/1/1865	23/11/1870	
Hong-Kong	21/12/1870	5/5/1872	
Pallaveram	20/5/1872	5/3/1873	
Madras	5/3/1873	25/11/1878	
Jubbulpur	3/12/1878	25/9/1879	
Meean Meer	29/9/1879	12/10/1880	
Deri Ghazi Khan	29/10/1879	28/12/1879	Right Wing onl
Jubbulpur	16/10/1880	16/10/1884	
Bellary	13/1/1885	7/12/1886	
On field service: Pyinmana	29/12/1886	4/4/1887	Detachments at nine outposts
Yemethin	12/4/1887	16/2/1888	
Pyawbwé	18/2/1888	1/4/1889	Detachment of 170 Rifles at Koni in the Shan States
Meiktila	1/4/1889	22/1/1890	Detachment of 2 companies at Pyinmana, and afterwards at Toungoo
Pallaveram	3/2/1890	25/3/1890	
Bangalore	26/3/1890	25/11/1892	Detachment of 40 men at Ootacamund
Cannanore	15/12/1892	31/11/1895	Detachments at Manjeri and Angadipuram
Thayetmyo	9/11/1895	8/12/1897	
Mindat Sakan	24/10/1895	-/1/1898	2 companies in cold weather, 1 company in the rains
Raipur	17/1/1898		Headquarters and Wing
Sambalpur	13/1/1898		One wing

PART IX

MISCELLANEOUS—PAY, PENSIONS, ETC.

IX

PAY

WHEN the regiment was first raised, the monthly pay of a subadar was 17 pagodas, of a jemadar 5 pagodas, of a havildar 3 pagodas, of a naick 2 pagodas 14 fanams, and of a sepoy 1 pagoda 29 fanams and 40 cash. As 80 cash were equivalent to 1 fanam, 42 fanams to 1 pagoda, which was itself worth Rs. $3\frac{1}{2}$, the pay of a sepoy was about Rs.6. The sepoy received 1 fanam and 30 cash per day as batta when employed in the field. 1776

In September 1784 the pay of recruit boys was fixed at 33 fanams (Rs.2, 12a.) per month. 1784

The pay of all ranks was increased in 1800, from which time the sepoy received 2 pagodas a month pay, and also 1 pagoda a day as batta when in the field, or in lieu thereof a seer of rice a day. 1800

On the 7th January 1818 it was ordered that all ranks should receive pay in the new currency of rupees, annas, and pies. 1818

In 1837 the pay of all ranks was again revised, the sepoys receiving Rs.7 a month, and Rs. $1\frac{1}{2}$ as extra batta when marching or in the field. Good conduct pay was simultaneously introduced for privates, consisting of an addition of one rupee to their pay after sixteen, and another rupee after twenty, years' service; and the Orders of British India (for a limited number of native officers) and of Merit (for non-commissioned officers and men) were created, with appropriate insignia and allowances. 1837

1846 During this year drummers were admitted to the benefit of the regulations regarding good conduct pay.

1877 In 1877 it was ordered that the sepoys should receive good conduct pay at the accelerated rates of one rupee after three, nine, and fifteen years' service.

1882 In 1882 good conduct pay was extended to havildars and naicks at the rate of one rupee after two, four, six, and eight years' service in the former grade, and at a similar rate after two and four years' service in the latter grade.

1886 In 1886 the periods at which privates could claim good conduct pay were reduced to three, six, and ten years. In the same year all ranks were granted free passages to and from their homes on furlough.

1895 From the 1st July 1895 an increase of Rs.2 per mensem to every non-commissioned officer and soldier in the regiment was sanctioned. Recruits enlisted after that date were also to receive the annual half-mounting allowance of Rs.5 from the date of enlistment, instead of from the date of completing eighteen months' service.

PENSIONS.

1786 In 1786 the pay of "invalids," that is, of men pensioned on account of wounds, was fixed at from 13 pagodas a month for subadars to 1 pagoda 29 fanams 40 cash for sepoys; and for "pensioners" at about half that sum. It was simultaneously ruled that no native soldier should be recommended for the invalid pension unless he had been wounded, or had served twenty years, and was fit for garrison duty.

1807 In 1807 ordinary pensions were revised, and settled at from 10 pagodas per month for subadars to 1 pagoda for sepoys.

In 1837 the pensions were divided into categories, the first obtainable after fifteen years' service, and the second after forty years' service, or earlier if disabled by wounds. These pensions ranged from Rs.25 under the first category to Rs.40 under the second for subadars, down to from Rs.4 to Rs.7 for privates. These pensions are still in force, with the exception that in 1877 native officers' pensions were increased, and the limit of service for the second, or "extraordinary," pension reduced from forty to thirty-two years. Even then, however, it is only permissible to sepoys of "unexceptionable" character; others, whose past fails to reach this elevated standard, having to be content with the inferior pension of Rs.4.

1837

1877

On the 19th September 1890, regulations were sanctioned by the Governor-General in Council for the employment of pensioned native soldiers in civil capacities, after retirement from the service, with the object of popularising military service amongst the desirable classes of India. It was sought to provide a ready means of communication between employers and pensioners who were desirous of employment. The receipt of pensions to be in no way affected by the acceptance of situations.

1890

On the 1st April 1892, with the sanction of the Right Honourable the Secretary of State for India, the Governor-General in Council directed that in future a step of honorary rank on retirement, with the title of honorary subadar-major, should be bestowed on such native officers as may be recommended by His Excellency the Commander-in-Chief as specially deserving of that honour. In very exceptional cases, officers who have served with special distinction, and who have attained the rank of subadar-major, may be granted the honorary rank of captain on retirement.

1892

MISCELLANEOUS.

1756 *Monthly musters* were ordered twenty years before the regiment was formed, with the object of discharging any men whom the commanding officers considered unfit for the service.

1759 The subject of *Pension to Heirs* of men killed, or who died on service, attracted the attention of Government at a very early date; and, in 1759, stoppages of one fanam a month from sepoys, up to one rupee from subadars, were made, wherefrom to provide a fund in case of casualties to the contributors. Three years later a step further in advance was essayed, and men volunteering for foreign service were informed "that such of them as are desirous of it, may have any part of their pay delivered to their families during their absence." This would appear to be the first mention of *Family* 1790 *Payments* in the history of the Madras Army. In 1790 fresh orders were issued regarding family payments 1797 before the outbreak of the Third Mysore War. In 1797 the "Right Honourable the President in Council," having reflected "with the highest satisfaction on the unexampled alacrity and spirit with which the Coast native troops had embarked for foreign service during the present war," was "naturally led to consider every means of preserving that spirit, and of rewarding their zeal." He was therefore pleased to resolve "a gratuity of three months' allowances to the relatives of all sepoys who had died or been killed 'to the eastward.'"

1804 In 1804, on the conclusion of the campaign in the Deccan, the half-pay of "every native officer and soldier" who had died or been killed was conferred upon his heir—this provision to last for twelve years if the deceased had sons, or "during the life of the nearest heir" in cases where the families consisted "only of women and aged persons."

Drums introduced.—In 1767 it was ordered that tom-toms should be discontinued in the native army "as soon as a sufficient number of men could be taught the different beats on the drum as practised in the European battalions." 1767

Regimental Registers were introduced on the 12th March 1770, "as the vacancies in sepoy battalions were chiefly occasioned by desertion, to prevent which, as much as possible, it was ordered that a book should be kept up for each battalion, in which was entered the name and description of every sepoy, setting forth his age, caste, district and village he came from, and where his family or relations reside." This Descriptive Register was revised in February 1785, since which time it has been in use in all native regiments. The Regimental Registers were abolished in 1897 on the introduction of the Sheet Roll. 1770

The Reading of the Indian Articles of War periodically on parade is a very old custom in the Madras Army, dating from 1st December 1774. 1774

Tents.—Up to October 1780 the sepoys provided their own tents when on service—although it is stated that European officers received them free—and were often compelled to use them in garrison, owing to insufficient hutting accommodation. In 1802, regiments received an allowance of 272 pagodas to supply their own tents, but this was abolished in 1808. A free issue was sanctioned, after the latter date, of single-poled tents, which pattern was in use until the two-poled tents were substituted on the 10th February 1817. General service tents were substituted for sepoy palls about 1885. 1780

Periodical Reliefs of Madras Regiments were instituted in January 1785, "as the continuance of sepoy battalions in the same district for a length of time had 1785 been followed by inconvenient results, more especially in the case of native troops in the Northern Circars, which had deserted in great numbers when required to march to the south during the war." Shortly afterwards, it was ordered that Madras Regiments should be relieved every six months. About 1820, reliefs were so arranged to bring back each regiment to the place where it had been chiefly raised, once in six years, so that the men should be able to visit their homes.

1805 *An Allowance for Officers' Mess* was sanctioned on the 30th June 1805, at the rate of 35 pagodas per mensem when marching or in the field, and of 18 pagodas when in cantonment.

1808 *Hutting Money.*—When Government supplied tents to the sepoys in 1808, they were granted hutting allowances; up to that time, accommodation for them had been provided by Commanding Officers.

1810 *Arrangements for Foreign Service.*—The period of absence of native troops was fixed at three years on the 9th March 1810, when they were also informed that the families of all fighting men and camp followers, who might either die or fall in battle during the absence of their corps from the Presidency, would be pensioned on the half-pay of their deceased relatives. The issue of woollen cloaks and pantaloons, as well as batta and rations, was also sanctioned.

1825 *The Eye Infirmary* at Madras, to which several men of the regiment have from time to time been sent, was established in 1825 for the reception of soldiers and others.

1866 The *Cart Fund* was introduced into the Madras Army on the 8th February 1866, "with the object of assisting in defraying the expenses attending the hire of carts from one station to another in course of relief."[1]

[1] Free carriage is not allowed to Madras Regiments on the line of march, but they draw marching batta instead.

Camp Colours of native infantry were ordered to be of the colour of the facings of the regiment, eighteen inches square and with number of regiment upon them, on the 5th February 1867. 1867

Running Drill was first introduced in the native army on the 12th April 1871. 1871

A *Band* was allowed in every regiment of native infantry in which the officers "expressed a desire to establish one, and in such cases the usual allowance would be paid in aid of its maintenance." (G.O.G., No. 32, of 1877.) 1877

COLONEL'S ALLOWANCES.

(*Extracts from the Madras Mail.*)

"The origin of the colonel's allowances dates from the first formation of standing armies, and consequently of the regiments that composed them, in Europe. The regiments were raised for the service of a monarch by a colonel, who received from the State Treasury an annual lump sum for providing clothing and equipments for his men. Out of this sum he recouped himself for his preliminary expenses, and made interest on the capital he had expended on the regiment, by the sale of commissions to officers, and by a profit on all the articles of clothing and equipment which he issued to the men; and also in a variety of other ways which would be rightly looked on as dishonest now, but which were considered quite legitimate in those times, or were at least tolerated by common consent. . . . Drawing pay for the full establishment while there were vacancies in the companies was another way of making an illegitimate profit, which was divided between the colonel and his captains. The former was the proprietor of the regiment, and unless he sold his interest in it, he continued to own it until his death; while, after his promotion to a general officer, or his retirement from active service, a lieutenant-colonel commanded the battalion of which it was composed. Regiments, long after they were raised, were known by the name of their proprietary colonels. As the modern organization of the State and Army gradually progressed, the proprietary functions of the colonel were by degrees transferred to the central administration; but the titular colonel still continued to draw a handsome profit from the contract for its clothing."

"The Indian colonel's allowances are the allowances of a colonel of a regiment of the East India Company's Army. For though, when that army was raised, the practice of the colonel's contracting for the supply of his regiment had been greatly modified in the British Army, yet such remnants of the practice as remained were adopted and continued in the Company's Army. The lieutenant-colonels in that army rose by seniority in succession to vacancies to be colonels of regiments, and to the emoluments attached to the position. The profits made on the regimental clothing contracts were called *off-reckonings.* The Government, when it first took the clothing of the troops into its own hands, out of the colonel's, lumped together the profit made on the contracts of each year, and divided it equally among the dispossessed colonels; but, afterwards, it was found convenient to commute these varying payments by a lump sum. The average yearly profit accruing for each regiment was found to be close on £700, and this amount was guaranteed to the regimental colonels for the future." After the amalgamation in 1865, when officers were promoted to colonels after a certain period of service, it was decided to admit everyone to off-reckonings after thirty-eight years' service. Officers admitted to the Staff Corps subsequently up to 1881 were allowed to obtain off-reckonings in the proportion of one to every thirty officers, but these allowances were abolished for all who joined after that date.

THE CLASSES FROM WHICH THE REGIMENT HAS BEEN RECRUITED SINCE 18

E. = ENLISTED. R. = RECEIVED FROM OTHER REGIMENTS.

		1845.		1846.		1847.		1848.		1849.		1850.		1851.		1852.		1853.		1854.	
		E.	R.	E.	R.	E.	R.	E.	R.	E.	R.	E	R.	E.	R.	E.	R.	E.	R.	E.	R.
Muhammedans.	Sayyid	6	...	5	...	...	...	2	...	1	...	2	1	3	...	7	...	4	...	1	...
	Shaik	13	1	22	...	...	...	11	1	13	2	6	...	10	1	8	1	12	...	17	...
	Moghal	1	...	3	...	...	...	2	...	4	...	5	...	1	...	2	...	...	...	...	...
	Pathan	1	...	7	...	...	...	...	2	1	1	2	...	2	...	2	...	2	...	1	...
	Moors	...	...	...	...	...	...	...	...	...	...	...	1	...	...	...	...	...	...	...	...
	Moplahs	...	...	...	...	...	...	...	...	...	...	...	...	...	...	...	...	...	...	...	...
	,, Sayyid	...	...	...	...	...	...	...	...	...	...	...	...	...	...	...	...	...	...	..	...
	,, Shaik	...	...	...	...	...	...	...	...	...	...	...	...	...	...	...	...	...	...	...	...
Tamilians.	Kavarai	2	...	...	...	...	...	...	...	...	...	...	...	...	...	...	...	...	...	...	...
	Vellaler	2	1	...	...	...	...	3	...	1	...	...	...	2	...	...	...	1	...	...	...
	Agambadians	1	...	1	...	...	...	...	...	...	...	...	...	...	...	...	1	...	...	...	...
	Malabar	...	1	3	...	...	...	...	1	...	...	3	...	1	...	3	...	4	...	8	...
	,, Goldsmiths	...	...	1	...	...	...	...	...	...	...	...	...	...	...	...	...	...	...	...	...
	Vannier	...	...	...	...	...	...	...	...	...	...	...	...	...	...	...	...	...	...	...	...
	,, Oil Monger	...	...	1	...	...	...	...	...	1	...	...	...	...	...	...	...	...	...	...	...
	Padiyachee	...	...	...	...	...	...	...	...	...	...	...	...	...	...	...	...	...	...	...	...
	Edayar	...	...	...	...	...	...	...	...	...	...	...	...	...	...	...	...	...	...	...	...
	Culler	...	...	...	...	...	...	1	...	...	...	...	...	1	...	...	...	...	...	...	...
	Reddy	...	...	...	...	...	...	...	...	...	...	...	...	...	...	...	...	...	...	...	...
	Moopanar	...	...	...	...	...	...	...	...	...	...	...	...	...	...	...	...	...	...	...	...
	Shanar	...	...	...	...	...	...	...	...	...	...	...	...	...	...	...	...	...	...	...	...
	Gonudon	...	...	...	...	...	...	...	...	...	...	...	...	...	...	...	...	...	...	...	...
	Malabar Dhobey	1	...	...	...	...	...	...	...	...	...	...	...	...	...	...	...	...	...	...	...
	,, Barber	...	...	...	1	...	...	...	...	...	...	...	...	...	...	...	...	...	...	...	...
	,, Natwa	...	...	...	...	...	...	...	...	...	...	...	...	...	...	...	...	...	...	...	...
Other Hindus.	Buljawar	3	...	6	...	...	...	...	...	2	...	5	...	...	...	2	...	...	...	...	...
	Gooval	1	...	...	...	...	...	...	...	...	...	...	...	...	...	...	...	...	...	...	...
	Gentoo	4	1	3	2	...	...	...	1	2	2	6	1	7	...	10	...	12	...	7	...
	Reddy Kulam	1	...	...	...	...	...	...	...	...	...	...	...	...	...	...	...	...	...	...	...
	Rachavar	...	...	...	...	...	...	...	...	...	...	...	...	...	...	...	...	...	...	...	...
	Shepherd (Golavar)	1	...	1	...	...	...	...	...	...	...	...	...	...	...	...	...	...	...	...	...
	Moottarasy	...	...	1	...	...	...	...	...	...	...	...	...	...	...	...	...	...	...	...	...
	Cummaravar (potter)	...	...	...	...	...	...	...	...	...	...	...	...	...	...	...	...	...	...	...	...
	Salivar (Weaver)	...	...	...	...	...	...	...	...	...	...	...	...	...	...	...	...	...	...	...	...
	Chakalavar (Washerman)	...	...	...	...	...	...	...	...	...	...	...	...	...	...	...	...	...	...	1	...
	Mangalavar	...	...	...	...	...	...	...	...	...	...	...	...	...	...	...	...	...	...	...	...
	Canarese	...	...	...	...	...	...	...	...	...	...	...	...	...	...	...	...	...	...	...	...
	Pullar	...	...	1	...	...	...	1	...	...	...	...	...	...	...	...	...	...	...	...	...
	Valloovar	...	...	1	...	...	...	...	...	...	...	...	...	...	...	...	...	...	...	...	...
	Pariah	3	...	...	...	...	...	...	...	2	...	1	...	1	...	...	...	2	...	1	...
	Chuckler	...	...	...	...	...	...	...	...	...	...	...	..	.	...	1	...	...	...	...	...
	Other caste	...	...	...	...	...	...	...	...	...	...	...	...	...	...	...	...	...	...	...	...
	Panachivon	...	...	...	...	...	...	...	...	...	...	...	...	...	...	...	...	...	...	...	...
	Rajput (Brahmins)	1	...	3	...	...	...	...	...	...	...	...	...	...	...	...	...	...	...	...	...
	Mahrattas	...	...	...	...	...	...	...	...	...	...	...	...	1	...	...	...	...	...	...	...
	Rajput Kshetre	...	...	1	...	...	...	...	...	...	...	...	...	...	...	...	...	...	...	...	...
	,, Shepherd	...	...	1	...	...	...	...	...	...	...	...	...	...	...	...	...	...	...	...	...
	Gomalbaudy	1	...	...	...	...	...	...	...	...	...	...	...	...	...	...	...	...	...	...	...
	Rajput Barber	...	...	...	...	...	...	...	...	...	...	...	..	...	...	...	...	...	1	...	...
	Rajput	5	...	4	...	...	...	1	...	...	...	2	...	3	...	1	...	...	...	...	...
Others.	Christian	2	...	1	...	2	...	...	...	6	...	4	...	3	...	...	...	3	...	3	...
	Indo-Briton	1	...	5	...	...	...	...	...	3	...	...	...	...	...	2	...	1	...	...	...
	East Indian	...	...	...	...	...	...	...	...	...	...	...	...	...	...	...	..	...	...	...	...
	Eurasian	...	...	...	...	...	...	...	...	...	...	...	...	...	...	...	...	...	...	...	...
	Portuguese	...	...	...	...	...	...	...	...	...	...	...	...	...	...	...	...	...	...	...	...
		50	4	71	3	2	...	21	5	36	5	36	3	35	1	38	2	41	1	39	...

1855.		1856.		1857.		1858.		1859.		1860.		1861.		1862.		1863.		1864.		1865.		1866.		1867.		1868.	
E.	R.	E.	R.	E.	R.	E.	R.	E.	R.	E.	R.	E.	R.	E.	R.	E.	R.	E.	R.	E.	R.	E.	R.	E.	R.	E.	R.
2	...	4	...	1	1	9	...	...	1	...	1	2	1	...	5	3	...	4	3	2	...	6	...	4	...	...	...
7	1	12	...	2	...	67	...	2	3	...	17	3	1	...	27	16	1	6	9	11	...	17	...	19	...	8	...
1	...	2	...	...	...	2	...	...	...	...	3	1	1	...	8	3	...	...	...	...	...	...	...	...	...	...	...
...	1	1	...	1	2	...	...	...	...	...	...	...	...	...	1	...	...	2	1	1	1	1	1	1	1	1	1
...	...	...	...	...	...	...	...	...	...	...	...	...	...	...	...	...	...	...	2	...	...	2	...	5	...	3	...
...	...	...	...	...	...	...	...	...	...	...	...	...	...	...	...	...	...	...	...	...	...	...	...	...	...	...	...
...	...	...	...	...	...	...	...	...	...	...	...	...	...	...	...	...	...	...	...	...	...	...	...	...	...	...	...
...	...	...	...	...	...	...	...	...	...	...	...	...	...	...	...	...	...	...	...	...	...	...	...	...	...	...	...
...	...	...	...	...	...	...	...	...	...	...	...	...	...	...	1	...	...	1	...	...	...	...	...	1	...	...	...
...	...	5	2	...	...	...	...	2	...	...	...	...	...	...	...	...	...	1	4	3	...	1	...	3	...	1	1
...	...	...	...	1	...	...	...	...	...	...	...	...	...	...	...	...	...	...	...	...	...	...	...	...	...	...	...
3	1	10	...	...	12	3	...	1	9	...	8	3	2	...	36	2	...	...	...	...	...	...	...	...	...	...	...
...	...	...	...	...	...	...	...	...	...	...	...	...	...	...	...	...	...	...	...	1	...	...	...	...	...	...	...
...	...	...	...	...	...	...	...	...	...	...	1	...	...	...	...	1	...	...	...	...	...	...	...	...	...	...	...
...	...	...	...	...	...	...	...	...	...	...	...	...	...	...	...	...	...	...	...	...	...	...	...	...	...	...	...
...	...	...	...	...	...	...	...	...	...	...	...	...	...	...	...	...	...	...	...	...	...	...	...	...	...	...	...
...	...	...	...	...	...	...	...	...	...	...	...	...	...	...	...	...	...	...	...	...	...	...	...	...	...	...	...
...	...	...	...	...	...	...	...	...	...	...	...	...	...	...	...	...	...	...	...	4	...	...	...	...	...	1	...
...	...	...	...	...	...	...	...	...	...	...	...	...	...	...	...	...	...	...	...	...	...	1	...	1	...	...	...
...	...	...	...	...	...	...	...	...	...	...	...	...	...	...	...	...	...	...	...	...	...	...	...	...	...	...	...
...	...	...	...	...	...	...	...	...	...	...	...	...	...	...	...	...	...	...	...	...	...	...	...	...	...	...	...
...	...	...	...	...	...	...	...	...	...	...	...	...	...	...	...	...	...	...	...	...	...	...	...	...	...	...	...
...	...	...	...	...	1	...	...	...	...	...	...	...	...	...	...	...	...	...	...	...	...	...	...	...	...	...	...
...	...	...	...	...	...	...	...	...	...	...	...	...	...	...	...	...	...	...	...	...	...	...	...	...	...	...	...
...	...	...	...	...	...	...	...	...	...	...	...	...	...	...	...	...	...	...	...	...	...	...	...	...	...	...	...
...	...	5	...	...	...	...	...	...	...	...	1	3	...	...	...	...	...	...	...	...	...	...	...	...	...	...	...
...	...	...	...	...	...	...	...	...	...	...	...	...	...	...	...	...	...	...	...	...	...	...	...	...	...	...	...
9	...	18	...	...	48	92	...	2	6	...	6	3	2	...	58	17	1	6	55	14	...	8	...	14	...	15	...
...	...	...	...	...	...	3	...	2	...	...	...	...	...	...	1	...	...	...	...	...	...	...	...	...	...	...	...
...	...	1	...	...	1	...	...	...	...	...	...	...	...	...	...	...	...	...	...	...	...	...	...	...	...	...	...
...	...	...	...	...	...	...	...	...	...	...	...	...	...	...	...	...	...	...	...	...	...	...	...	...	...	...	...
...	...	...	...	...	...	...	...	...	...	...	...	...	...	...	1	...	...	...	...	...	...	...	...	...	...	...	...
...	...	...	...	...	...	2	...	...	...	...	...	...	...	...	...	...	...	...	...	...	...	...	...	...	...	...	...
...	...	...	...	...	...	6	...	...	...	...	...	...	...	...	...	...	...	...	...	...	...	...	...	...	...	...	...
...	...	...	...	...	...	4	...	...	...	...	...	...	...	...	...	...	...	...	...	...	...	...	...	...	...	...	...
...	...	...	...	...	...	1	...	...	...	...	...	...	...	...	...	...	...	...	...	...	...	...	...	...	...	...	...
...	...	...	...	...	...	...	...	...	...	...	...	...	...	...	1	...	...	...	...	...	...	...	...	...	...	...	...
...	1	...	...	...	1	1	...	...	...	...	2	...	...	...	1	...	...	...	...	...	...	...	...	...	...	1	...
1	...	3	...	...	...	4	...	...	...	...	...	...	...	...	...	...	...	...	...	...	...	2	...	2	...	1	...
1	...	...	...	...	...	10	...	...	5	...	...	...	...	...	13	...	...	...	2	...	...	1	...	2	...	...	...
...	...	...	...	...	1	...	...	...	...	...	...	...	...	...	...	...	...	...	...	...	...	...	...	...	...	...	...
...	...	...	...	...	...	...	...	...	...	...	...	1	...	...	3	...	...	...	...	...	...	...	...	...	...	...	...
...	...	...	...	...	...	...	...	...	...	...	...	...	...	...	2	...	...	...	...	...	...	...	...	...	...	...	...
...	...	...	...	...	1	...	...	...	...	...	...	...	3	...	...	...	...	...	...	...	...	...	...	...	...	...	...
...	...	...	...	...	...	...	...	...	...	...	...	...	...	...	...	...	...	...	...	...	...	...	...	...	...	...	...
...	...	...	...	...	...	...	...	...	...	...	...	...	...	...	...	...	...	...	2	2	...	...	...	...	...	...	...
...	...	1	...	...	...	...	...	...	...	...	...	...	...	...	...	...	...	...	3	6	...	...	...	4	...	...	...
...	...	...	...	...	...	...	...	...	...	...	...	...	...	...	...	...	...	...	...	...	...	...	...	...	...	...	...
...	...	...	...	...	...	...	...	...	...	...	...	...	...	...	...	...	...	...	...	...	...	...	...	...	...	...	...
...	...	3	1	...	2	1	...	...	...	...	...	...	...	...	6	...	...	...	1	...	...	...	...	...	...	...	...
1	...	2	...	...	...	14	...	...	...	...	...	...	...	...	9	3	...	3	...	...	...	...	...	...	...	...	...
...	...	...	...	...	...	...	...	...	...	1	...	...	...	...	1	...	...	...	...	...	...	...	...	...	...	...	...
...	...	...	...	...	...	...	...	...	...	...	...	...	...	...	...	...	...	...	...	...	...	...	...	...	...	...	...
...	...	...	...	...	...	...	...	...	...	...	...	...	...	...	...	...	...	...	...	...	...	...	...	...	...	...	...
...	...	...	...	...	...	...	...	...	...	...	...	...	...	...	...	...	...	...	...	...	...	...	...	...	...	...	...
25	4	67	3	5	70	219	...	9	24	1	39	16	10	...	174	45	2	23	82	44	1	39	1	56	1	31	2

1869.		1870.		1871.		1872.		1873.		1874.		1875.		1876.		1877.		1878.		1879.		1880.		1881.		1882.		1883.	
E.	R.	E.	R.	E.	R.	E.	R.	E.	R.	E.	R.	E.	R.	E.	R.	E.	R.	E.	R.	E.	R	E.	R.	E.	R.	E.	R.	E.	
4	...	2	10	...	...	...	...	...	...	1	...	...	...	1	...	2	...	2	...	...	...	2	...	2	...	1	1	6	...
15	...	9	20	7	...	1	...	1	...	1	...	11	...	10	1	7	...	8	2	9	...	17	...	25	...	13	7	13	...
...	...	...	...	...	...	...	...	...	...	...	...	1	...	...	...	...	...	...	...	...	...	...	...	3	...	...	1	...	...
...	...	1	5	1	...	...	...	...	...	...	...	3	...	2	...	2	...	...	...	1	...	...	...	1	...	...	1	2	...
2	...	...	...	...	...	...	...	...	...	...	...	...	...	...	...	...	...	...	...	...	...	...	...	...	...	...	...	...	...
...	...	...	...	...	...	...	...	...	...	...	...	...	...	...	...	...	...	...	...	...	...	...	...	...	...	...	...	...	...
...	...	...	...	...	...	...	...	...	...	...	...	...	...	...	...	...	...	...	...	...	...	...	...	...	...	...	...	...	...
...	...	...	...	...	...	...	...	...	...	...	...	...	...	...	...	...	...	...	...	...	...	...	...	...	...	...	...	...	...
...	...	...	...	...	...	...	...	...	...	...	...	...	...	...	...	...	...	...	...	...	...	...	...	...	...	...	...	...	...
5	...	...	1	...	...	...	...	...	...	...	...	...	...	...	...	...	...	...	...	...	...	...	...	...	...	...	...	1	...
...	...	...	...	...	...	...	...	...	...	...	...	...	...	...	...	...	...	...	...	...	...	...	...	...	...	...	...	...	...
...	...	2	25	1	...	...	...	...	...	1	...	2	...	3	1	6	...	2	...	2	...	7	...	1	...	12	...	8	...
...	...	...	...	...	...	...	...	...	...	...	...	...	...	...	...	...	...	...	...	...	...	...	...	...	...	...	...	...	...
...	...	...	...	...	...	...	...	...	...	...	...	...	...	...	...	...	...	...	...	...	...	...	...	...	...	...	...	1	...
...	...	...	...	...	...	...	...	...	...	...	...	...	...	...	...	...	...	...	...	1	...	...	...	...	...	...	...	2	...
...	...	...	...	...	...	...	...	...	...	...	...	...	...	...	...	...	...	...	...	...	...	...	...	...	...	...	...	...	...
...	...	...	...	...	...	...	...	...	...	...	...	...	...	...	...	...	...	...	...	...	...	...	...	...	...	...	...	...	...
1	...	...	...	...	...	...	...	...	...	...	...	...	...	...	...	...	...	...	...	...	...	...	...	...	...	...	...	...	...
...	...	...	...	...	...	...	...	...	...	...	...	...	...	...	...	...	...	...	...	...	...	...	...	...	...	...	...	...	...
...	...	...	...	...	...	...	...	...	...	...	...	...	...	...	...	...	...	...	...	...	...	...	...	...	...	...	...	...	...
...	...	...	...	...	...	...	...	...	...	...	...	...	...	...	...	...	...	...	...	...	...	...	...	...	...	...	...	...	...
...	...	...	1	...	...	...	...	...	...	...	...	...	...	...	...	...	...	...	...	...	...	...	...	...	...	...	...	...	...
1	...	...	...	...	...	...	...	...	...	...	...	...	...	...	...	...	...	...	...	...	...	...	...	...	...	...	...	...	...
...	...	...	1	...	...	...	...	...	...	...	...	...	...	...	...	...	...	...	...	...	...	...	...	...	...	...	...	...	...
...	...	...	...	...	...	...	...	...	...	...	...	...	...	...	...	...	...	...	...	...	...	...	...	...	...	...	...	...	...
...	...	12	45	6	...	1	...	1	...	2	...	13	...	11	...	24	...	15	1	26	...	38	...	19	...	25	27	45	2
...	...	...	...	...	...	...	...	...	...	...	...	...	...	...	...	...	...	...	...	...	...	...	...	...	...	...	...	...	...
20	8	...	2	...	...	...	...	...	...	...	...	...	...	...	...	...	...	...	...	...	...	...	...	...	...	...	1	...	...
...	...	...	...	...	...	...	...	...	...	...	...	...	...	...	...	...	...	...	...	...	...	...	...	...	...	...	...	...	...
...	...	...	...	...	...	...	...	...	...	...	...	...	...	...	...	...	...	...	...	...	...	...	...	...	...	...	...	...	...
...	...	...	...	...	...	...	...	...	...	...	...	...	...	...	...	...	...	...	...	...	...	...	...	...	...	...	...	...	...
...	...	...	...	...	...	...	...	...	...	...	...	...	...	...	...	...	...	...	...	...	...	...	...	...	...	...	...	...	...
...	...	...	...	...	...	...	...	...	...	...	...	...	...	...	...	...	...	...	...	...	...	...	...	...	...	...	...	...	...
...	...	...	...	...	...	...	...	...	...	...	...	...	...	...	...	...	...	...	...	...	...	...	...	...	...	...	...	...	...
...	...	...	...	...	...	...	...	...	...	...	...	...	...	...	...	...	...	...	...	...	...	...	...	...	...	...	...	...	...
...	...	...	1	...	...	...	...	...	...	...	...	...	...	...	...	...	...	...	...	...	...	...	...	...	...	...	...	...	...
...	...	...	...	...	...	...	...	...	...	...	...	...	...	...	...	...	...	...	...	...	...	...	...	...	...	...	...	1	...
2	...	...	...	...	...	...	...	...	...	...	...	...	...	...	...	...	...	...	...	...	...	...	...	...	...	...	...	...	...
2	...	...	...	...	...	...	...	...	...	...	...	...	...	...	...	...	...	...	...	...	...	...	...	...	...	...	...	3	...
...	...	1	7	1	...	1	...	...	...	...	...	...	...	...	...	1	...	...	...	1	...	4	1	4	...	...	...	...	...
...	...	...	...	...	...	...	...	...	...	...	...	...	...	...	...	...	...	...	...	...	...	...	...	...	...	...	...	...	...
...	...	...	4	...	...	...	...	...	...	...	...	...	...	...	...	...	...	...	...	...	...	...	...	...	...	...	...	...	...
...	...	...	...	...	...	...	...	...	...	...	...	...	...	...	...	...	...	...	...	...	...	...	...	...	...	...	...	...	...
1	...	...	6	...	...	...	...	...	...	...	...	...	...	...	...	...	...	...	...	...	...	...	...	...	...	...	...	...	...
...	...	...	...	...	...	...	...	...	...	...	...	...	...	...	...	...	...	...	...	...	...	...	...	...	...	...	...	...	...
...	...	...	...	...	...	...	...	...	...	...	...	...	...	...	...	...	...	...	...	...	...	...	...	...	...	...	...	...	...
...	...	...	...	...	...	...	...	...	...	...	...	...	...	...	...	...	...	...	...	...	...	...	...	...	...	...	...	...	...
...	...	...	...	...	...	...	...	...	...	...	...	...	...	...	...	...	...	...	...	...	...	...	...	...	...	...	...	...	...
...	...	...	...	...	...	...	...	...	...	...	...	...	...	...	...	...	...	...	...	...	...	...	...	...	...	...	...	...	...
...	...	...	...	...	...	...	...	...	...	...	...	...	...	...	...	...	...	2	...	1	...	3	...	8	...	1	...	1	...
...	...	4	12	1	...	1	...	1	...	...	...	...	...	1	...	4	...	1	...	...	...	3	1	1	...	...	...	4	...
...	...	1	1	...	...	...	...	...	...	2	...	2	...	...	...	2	...	3	...	1	...	...	...	...	...	1	...	1	...
...	...	...	1	...	...	...	...	...	...	...	...	...	...	...	...	...	...	...	...	...	...	...	...	...	...	...	...	...	...
...	...	...	...	...	...	...	...	...	...	...	...	...	...	...	...	...	...	...	...	...	...	...	...	...	...	...	...	...	...
...	...	...	...	...	...	...	...	...	...	...	...	...	...	...	...	...	...	...	...	...	...	...	...	...	...	...	...	...	...
53	8	32	142	17	...	4	...	3	...	7	...	32	...	28	2	48	...	33	3	42	...	74	2	64	...	53	38	88	2

1884.		1885.		1886.		1887.		1888.		1889.		1890.		1891.		1892.		1893.		1894.		1895.		1896.		1897.	
E.	R.	E.	R.	E.	R.	E.	R.	E.	R.	E.	R.	E.	R.	E.	R.	E.	R.	E.	R.	E.	R.	E.	R.	E.	R.	E.	R.
...	...	18	...	6	...	...	...	5	...	3	...	6	8	10	4	3	4	2	1	1	...	4	1	1	...	6	...
2	...	44	1	28	2	3	4	13	2	5	...	22	27	48	10	5	13	5	7	18	...	15	3	10	2	32	...
...	...	3	...	1	...	...	...	...	...	...	...	...	1	3	...	2	1	2	1	4	...	1	3	1	1	2	...
...	...	3	1	9	1	...	...	...	...	1	...	2	5	7	2	...	4	4	...	3	1	1	...	3	...	8	...
...	...	...	...	...	...	...	...	...	...	...	...	...	...	...	...	...	...	...	...	...	...	...	...	...	...	...	...
...	...	...	...	...	...	...	...	...	...	...	...	...	...	...	...	...	...	...	...	...	...	...	...	...	...	...	...
...	...	...	...	1	...	...	...	...	...	...	...	...	...	...	...	...	...	...	...	...	...	...	...	...	...	...	...
...	...	...	...	7	1	...	...	...	...	...	...	...	...	...	...	...	...	...	...	...	...	...	...	...	...	...	...
...	...	...	...	3	...	...	...	1	...	5	...	...	...	...	...	...	...	...	...	...	...	...	...	1	1	4	...
...	...	...	...	...	...	...	...	...	...	1	...	2	1	3	5	1	4	3	...	24	...	6	7	...	...	1	...
...	...	...	...	...	...	...	...	2	...	...	...	...	...	...	...	...	1	1	...	...	2	...	...	...	...	1	...
1	...	1	1	4	...	2	3	3	...	7	...	4	10	1	5	...	...	...	4	...	...	...	...	...	...	...	...
...	...	...	...	...	...	...	...	...	...	...	...	...	...	...	...	...	...	...	...	...	...	...	...	...	...	...	...
...	...	...	...	...	...	...	...	1	...	1	...	2	3	1	2	1	...	...	...	3	...	4	...	3	1	3	...
...	...	...	...	...	...	...	...	...	...	...	...	...	...	...	...	...	...	...	...	...	...	...	...	...	...	...	...
...	...	...	...	...	...	...	...	...	...	...	...	...	...	...	...	...	...	...	...	...	...	...	...	1	...	...	...
...	...	...	...	...	...	...	...	...	...	...	...	...	...	...	...	...	...	...	...	1	...	...	...	3	...	2	...
...	...	...	...	...	...	...	...	...	...	...	...	...	...	...	...	...	...	...	...	2	...	...	2	2	...	2	...
...	...	...	...	...	...	...	...	3	...	...	...	2	3	...	3	...	...	...	...	...	...	...	...	...	...	...	...
...	...	...	...	...	...	...	...	...	...	...	...	...	...	...	...	...	...	...	...	1	...	1	...	...	...	...	...
...	...	...	...	...	...	...	...	...	...	...	...	...	...	...	...	...	...	...	...	...	...	2	...	...	...	...	...
...	...	...	...	1	...	...	...	1	...	2	...	...	...	...	...	...	...	...	...	...	...	...	...	1	...	...	...
...	...	1	...	...	...	...	1	...	...	...	...	...	...	...	...	...	...	...	...	...	...	...	...	1	...	...	...
...	...	...	...	...	...	...	...	...	...	...	...	...	...	...	...	...	...	...	...	...	...	...	...	...	...	...	...
...	...	...	...	...	...	...	...	...	...	...	...	...	...	...	...	...	...	...	...	...	...	1	...	...	...	1	...
[illegible]	...	31	...	7	1	2	27	15	...	7	...	2	17	2	12	...	...	...	...	...	...	...	...	...	...	...	...
...	...	...	...	...	...	...	...	...	...	...	...	...	...	...	...	...	...	...	...	...	...	...	...	...	...	...	...
...	...	...	...	...	...	...	...	...	...	...	...	...	...	1	1	...	25	2	1	...	...	1	...	...	...	...	...
...	...	...	...	...	...	...	...	...	...	...	...	...	...	...	...	...	...	...	...	...	...	...	...	...	...	...	...
...	...	...	...	...	...	...	...	...	...	...	...	...	...	...	...	...	...	...	...	...	...	...	...	...	...	...	...
...	...	...	...	...	...	...	...	...	...	...	...	...	...	...	...	...	...	...	...	...	...	...	...	...	...	...	...
...	...	...	...	...	...	...	...	1	...	...	...	...	...	...	...	...	...	...	...	...	...	...	...	...	...	1	...
...	...	...	...	...	...	...	...	...	...	...	...	...	...	...	...	...	...	...	...	...	...	...	...	...	...	...	...
...	...	...	...	...	...	...	...	...	...	...	...	...	...	...	...	...	...	...	...	...	...	...	...	...	...	...	...
...	...	...	...	...	...	...	...	...	...	...	...	...	...	...	...	...	...	...	...	...	...	...	...	...	...	...	...
...	...	...	...	...	...	...	...	...	...	...	...	...	...	...	...	...	...	...	...	...	...	...	...	...	...	...	...
...	...	...	...	...	...	...	...	...	...	...	...	...	1	...	...	...	...	...	...	...	2	...	...	...	...	...	...
...	...	...	...	...	...	...	...	...	...	1	...	...	...	...	...	...	...	...	...	...	1	...	...	...	...	...	...
...	...	...	...	...	...	...	...	...	...	1	...	...	...	...	...	...	...	...	...	...	...	...	...	...	...	...	...
...	...	...	...	2	1	...	...	...	...	1	...	...	1	...	2	2	1	...	1	2	...	3	...	...	1	...	...
...	...	...	...	...	...	...	...	...	...	...	...	...	...	...	...	...	...	...	...	...	...	...	...	...	...	...	...
...	...	2	...	...	...	...	...	3	...	1	...	2	6	3	4	...	...	1	...	...	...	1	...	...	...	2	...
...	...	...	...	2	...	...	...	...	...	...	...	...	...	...	...	...	...	...	...	...	...	...	...	...	...	...	...
...	...	...	...	...	...	...	...	...	...	...	...	...	...	...	1	...	...	...	...	...	...	...	...	...	...	...	...
...	...	...	...	...	...	...	...	...	...	...	...	...	...	2	2	...	...	...	...	3	...	...	1	1	...	1	...
...	...	...	...	...	...	...	...	...	...	...	...	...	...	...	...	...	...	...	...	...	...	...	...	...	...	...	...
...	...	...	...	...	...	...	...	...	...	...	...	...	...	...	...	...	...	...	...	...	...	...	...	...	...	...	...
...	...	...	...	...	...	...	...	...	...	...	...	...	...	...	...	...	...	...	...	...	...	...	...	...	...	...	...
...	...	...	...	...	...	...	...	...	...	...	...	...	...	...	...	...	...	...	...	...	...	...	...	...	...	...	...
[illegible]	...	...	...	1	1	...	2	4	...	1	...	1	...	8	...	...	...	...	...	...	...	...	...	2	...	...	...
[illegible]	...	...	...	3	1	2	7	1	...	6	...	3	10	12	3	...	...	1	...	3	...	2	...	...	...	12	...
[illegible]	...	1	...	1	...	...	...	...	...	1	...	...	...	...	...	...	...	1	...	...	1	...	...	...	...	...	...
...	...	8	...	...	...	...	...	...	...	...	...	...	...	1	...	1	...	...	...	...	...	1	...	...	...	1	...
...	...	...	...	...	...	...	...	...	...	...	...	...	...	...	...	...	...	1	1	...	...	1	...	...	...	...	...
...	...	...	...	1	...	...	...	...	...	...	...	...	...	...	...	...	...	...	...	...	...	...	...	...	...	...	...
[illegible]	...	112	3	77	8	9	44	53	2	44	...	48	93	102	56	15	53	23	16	65	7	44	17	30	6	79	76

QUEEN'S AND REGIMENTAL COLOURS,

PLATE I.

NATIVE OFFICERS, NON-COMMISSIONED OFFICERS, AND SEPOYS, OF MADRAS ARTILLERY AND INFANTRY.
1791.

See page 201.

PLATE II.

SUBADAR, GRENADIER, AND RECRUIT OF THE MADRAS NATIVE INFANTRY, 1799.
(From Gold's Oriental Drawings.)

See page 201.

SEPOY,
1819.

BRITISH OFFICER WITH GORGET AND FUSIL AT THE "CARRY" MARCHING PAST,
1776.

See page 200.

PLATE IV.

BRITISH OFFICER OF NATIVE INFANTRY LEADING HIS MEN,
1816.

See page 204.

PLATE V.

SEPOY, 1806. *See page 204.*

SEPOY, 1824. *See page* [illegible]

BANDSMAN, 1840. *See page 210.*

SUBADAR, 1850. *See page 210.*

PLATE VI.

DRILL ORDER, 1860. SEPOY.

See page 213.

REVIEW ORDER, 1860. SEPOY.

See page 213.

DRILI ORDER, 1861. SEPOY.

See page 213.

RECRUIT BOY, 1866.

See page 213.

Regimental Badge,
1850.

Regimental Crest,
1850.

Regimental Badge,
1865.

Regimental Crest,
1865.

Regimental Crest,
1865.

Regimental Badge
and Crest,
1891.

Regimental Crest
on Notepaper,
1896.

1.

www.ingramcontent.com/pod-product-compliance
Ingram Content Group UK Ltd.
Pitfield, Milton Keynes, MK11 3LW, UK
UKHW021840270726
14058UKWH00002B/252

9 781845 743284